BEATING THE ODDS ON THE NORTH PACIFIC

A Guide to Fishing Safety

Edited by Susan Clark Jensen

AMSEA

Alaska Marine Safety Education Association
Box 2592
Sitka, AK 99835
(907) 747-3287

Sea Grant

University of Alaska Sea Grant
P.O. Box 755040 • Fairbanks, AK 99775-5040
(907) 474-6707 • FYPUBS@uaf.edu
http://www.uaf.edu/seagrant

MAB-41
Third Edition 1998   Price: $12.00

Elmer E. Rasmuson Library Cataloging-in-Publication Data

Beating the odds on the North Pacific : a guide to fishing safety /
    edited by Susan Clark Jensen.
3rd. ed.
(MAB-41), 1998
    1. Fishing boats—Safety measures. 2. Fisheries—Safety measures. 3. Fisheries—
Accidents and injuries. I. Jensen, Susan Clark.
II. Alaska Sea Grant College Program. III. Alaska Sea Grant Marine Advisory Program.
IV. Title. V. Series: Marine advisory bulletin ; 41.

SH343.9.B43 1998

ISBN 1-56612-054-3

About the Editor

Editor Susan Clark Jensen lives in Sitka, Alaska, where she trains safety instructors and teaches for the Alaska Marine Safety Education Association (AMSEA). Jensen was a founding board member of AMSEA, and has served as Emergency Medical Services coordinator for both Southeast Alaska Regional Health Corporation and Southeast Region EMS Council.

Credits

Cover photo is by Jim Nilsen, cover design is by Susan Burroughs, copy editing is by Barbara Matthews, and desktop publishing is by Lisa Sporleder and Carol Kaynor. Sue Keller, Kurt Byers, Lisa Sporleder, and Carol Kaynor also made editorial contributions.

Photos: K. Byers p. 12, 46, 68; J. Dzugan p. 184; A. French p. 118; D. Garza p. vi; J. Graves, *Kodiak Daily Mirror* 164; *Kodiak Daily Mirror* p. 38, 76; Kodiak Fire Dept. p. 148; D. Mercy p. 28, 182; North Star Survival p. 102. Illustrations: K. Beebee p. 137; V. Culp p. 55, 57-60, 72, 106; S. Lawrie p. 20, 30, 32, 43, 44, 52, 53, 55, 56, 61, 62, 65, 66, 71; K. Lundquist p. 21, 31 and 40 (after B. Bulloch, *Pacific Fishing*), 35 (after W. Tupper), 54, 83, 114, 126, 131, 141; J. Schmitts p. 4, 15, 30, 48, 50, 51, 64, 84-95, 111, 150-155, 171; L. Sporleder p. 14, 23, 24, 166, 177, 178 (after USCG); W. Tupper, *Fishermen's News* p. 159, 175; *The First Aid Book*, Southeast Alaska Regional Health Corp., Sitka, 1989 p. 105, 125, 137; *Healthcare Provider's Manual for Basic Life Support*, 1988, © American Heart Association, reproduced with permission, p. 120-123, 127, 128; K. Sherrodd p. 80, 81, 155, 189, 204, 206.

AMSEA provided support with help from an injury prevention grant from the Centers for Disease Control, U.S. Dept. of Health and Human Services.

This manual was produced by the Alaska Sea Grant College Program, cooperatively supported by the U.S. Dept. of Commerce, NOAA Office of Sea Grant, grant no. NA90AA-D-SG066, projects A/71-01 and A/75-01; and by the University of Alaska with state funds. The University of Alaska is an affirmative action/equal opportunity employer and educational institution.

Sea Grant is a unique partnership with public and private sectors combining research, education, and technology transfer for public service. This national network of universities meets changing environmental and economic needs of people in our coastal, ocean, and Great Lakes regions.

Table of Contents

Acknowledgments

Many people helped immensely by reviewing this manuscript in its various stages. My thanks to John Doyle, Dolly Garza, Jim Herbert, Bob Jacobsen, Dug Jensen, Norm Lemley, Hank Pennington, Glenn C. Sicks, and Chris Volkle for their comprehensive comments. Ken Coffland, Ed Diener, Dave Glamann, Bob Kuenzel, Stephen Lawrie, John Manning, Dave McKinney, Bruce Merchant, Patty Owen, Howie Pitts, Tom Schornak, Laddie Shaw, Al Snelling, Steph Steffen, Wade Watkins, and Tom Wood, M.D., also reviewed portions of the book. I am especially grateful to Jerry Dzugan at the Alaska Marine Safety Education Association for his advice and extensive editing, support, and encouragement over the several years of this project.

The artists listed in the credits section deserve much credit for the patience and creativity they showed in their illustrations. I am much obliged to the Sea Grant Marine Advisory Program staff in Sitka, Alaska, and the Sea Grant folks in Fairbanks. Original text contributors include Paula Cullenberg, Jerry Dzugan, Andrew M. Embick, M.D., Dolly Garza, Jim Herbert, Dave McKinney, Hank Pennington, Tom Schornak, Glenn C. Sicks, Rick Steiner, and Chris Volkle.

Chapter IV was adapted from an article by Rick Steiner in the Marine Advisory Program *Alaska Marine Resource Quarterly*, Vol. II, No. 1. The tendonitis and carpal-tunnel syndrome sections were adapted from *Tendonitis and Related Afflictions in Fishermen and Processing Workers* by Rick Steiner and Andrew M. Embick, M.D., Alaska Sea Grant ASG No. 26. Many thanks to A.K. Larssen, who wrote *Safety Notes for the North Pacific Fisherman,* Alaska Sea Grant MAB No. 3. Thanks to Jon Kjaerulff, Jerry Dzugan, Dug Jensen, Hank Pennington, and Glenn Sicks, who developed the four suggested drills in Chapter XII.

Information about personal survival kits (Chapter I), much of the information about personal flotation devices (Chapter V), and the section on paralytic shellfish poisoning in Chapter VIII came from AMSEA's *Marine Safety Instructor Training Manual.* AMSEA also provided information for Chapter XII and the book reviews in the resources section, and paid for illustrations. Information in Chapter X was adapted from USCG Navigation and Inspection Circular No. 5-86.

Throughout the book male pronouns are used; they are intended to apply to both males and females.

Susan Clark Jensen, Editor

Foreword

Safety is a personal responsibility that we all share. This is particularly true for the North Pacific fishing industry. Vessel owners and skippers, of course, have a responsibility to provide safe working conditions, but ultimately safety must be an individual concern. Gaining an awareness for the hazardous working conditions in the North Pacific is not easy. Experienced fishermen, with generations of family knowledge of the sea behind them, still die each year. No one is perfectly safe at sea! Some, however, substantially improve their odds of surviving by being prepared before leaving the dock.

Seaworthy vessels with experienced skippers and crew catch more fish and have a much greater chance of returning home safely. These boats are in demand, and they choose the most experienced deckhands. Unfortunately, the opposite is also true—unseaworthy vessels with inexperienced skippers hire newcomers with no commercial fishing experience. Homework is necessary for the new fishermen to improve their chances for hire on good boats and to help them recognize the unseaworthy vessels that prey on the uninitiated. This book reaches out to not only the veteran fisherman, but to the new deckhand as well with an easy to understand message for improving personal and vessel safety on a commercial fishing vessel.

Concern for the safety of North Pacific fishermen evolves from exceptionally high losses each year. For waters off Alaska alone, the average loss of lives and vessels for 1987–1991 was 36 lives and 42 vessels per year. Injury rates are believed to be significantly higher, but accurate numbers are difficult to obtain.

The Commercial Fishing Industry Vessel Safety Act of 1988 mandated that commercial fishing vessels carry minimum safety equipment, and required some vessels to conduct safety instruction and drills each month. Survival craft, immersion suits, EPIRBs, and visual distress signals are among the significant new equipment requirements. A checklist of these new requirements, as well as pre-existing requirements, is available from the Coast Guard by calling (907) 463-2212 (800-478-7369 in Alaska). All fishermen, regardless of experience, should ensure that their vessels meet at least these minimum requirements. Familiarity with this book and Coast Guard requirements is basic homework for all fishermen regardless of experience. It will aid you in personally recognizing and correcting unsafe conditions that could take your vessel and your life.

> Glenn C. Sicks, Lieutenant Commander, U.S. Coast Guard
> Seventeenth District Fishing Vessel Safety Coordinator

Preparing for a Dangerous Job

At 7:24 a.m. on November 15, 1985, the F/V *Lasseigne* was in trouble 20 miles off Siletz Bay, Oregon. She had a bad list and her captain was unable to get to the fish hold to tell where the water was coming from. He radioed the Coast Guard for assistance and was told to make sure the crew had their life jackets on. They did.

At 8:38 a.m. a U.S. Coast Guard helicopter crew sighted the now-capsized 73-foot trawler. No one is sure what happened during the 59 minutes after the distress call. When the Coast Guard arrived, all three men aboard were dead. Two were found floating in their life jackets. The third—whose body was never recovered—was presumed to have been trapped in the vessel. When the ship sank, it had on board one survival suit—which no one used. It had no inflatable life raft.

Families of the three men sued the vessel's owner. On May 8, 1987, Judge Edward Leavy, U.S. District Court, found the *Lasseigne* unseaworthy with the privity and knowledge of its owners. Three of the areas specifically cited by Judge Leavy follow:

1) **Fishing an Unstable Vessel**: Conversion of the *Lasseigne* from a shrimper to a trawler added nets, steel net reels and trawl doors, and a steel hydraulic winch that weighed more than half a ton and was mounted on the mast. Stability tests were never done after the conversion.

2) **Uncorrected Problems**: Known difficulties with the bilge pump and bilge alarm had not been fixed.

3) **Lack of Lifesaving Equipment**: Judge Leavy stated, "I find as a matter of law that the lack of a suitable life raft and survival suits for each crew member rendered this vessel unseaworthy." Also noted was the failure to maintain a watch at night.

The point about lifesaving equipment is especially significant. When the case was heard, life rafts and survival suits were **not** required by law or regulation, but the decision sets a precedent that holds vessel owners legally responsible for having adequate safety equipment on board their vessel. As Judge Leavy noted in his decision, "The owner of a vessel has an absolute and non-delegable duty to provide a seaworthy ship."

Having adequate survival equipment doesn't just make good legal sense—it can and does save lives. It is essential in one of the most dangerous industries in the United States.

Proper preparation begins with your care of the vessel, and includes survival equipment, training, on board drills, and having the will to live.

Vessel

Your fishing vessel is more than a vehicle to get to and from the fishing grounds and a storage place for your catch. It's also your temporary home—your shelter from the ocean and elements. Lose it, and you are instantly in danger.

Good skippers and deckhands know that it takes constant attention to keep a fishing vessel in top shape. Equipment, engines, electrical systems, and gear must be regularly inspected and maintained. Faulty through-hull fittings, broken bilge pumps, and unrepaired alarms can contribute to a sinking quicker than we'd like to think.

It is important to anticipate and prevent as many potential problems as possible. One of the best ways to ensure this is by doing a prevoyage check, much as aircraft pilots perform before they take off. The frequency of inspection and repair depends on the type of vessel, its equipment, and the fishery. Consult the resources section for more information on vessel inspection and maintenance. Keep in mind that U.S. Coast Guard safety requirements are for minimal safe conditions only. Most vessels need more safety features to make fishing conditions safe.

If something does go wrong with your vessel and you need to abandon ship, having well-maintained survival gear and a crew that knows how to use it can greatly increase your chances of survival.

Survival Equipment

No fisherman likes to think he will need to use survival equipment, but a life raft, survival (immersion) suit, emergency signals, personal survival kit, and training can help save your life. Survival equipment that helps protect you from the cold water, wind, rain, and snow, and allows you to signal for assistance will help keep you alive until you are rescued.

Regular maintenance is worth the time. Poor maintenance can mean disaster later. (K. Byers photo)

Life Rafts

Do life rafts help save lives? Just ask the survivors from the F/Vs *Tidings, Mary L, Unimak, Rebecca, Trilby,* and *Ranger,* to name a few. Life rafts were a key factor in their survival.

Life rafts work because they shelter you from the environment. Although all rafts are configured a bit differently, they all share some common features.

When considering which raft to buy, compare brands, ask other fishermen which raft they recommend and why, and talk to people who have survived in a raft. Some marine safety and survival training courses offer the opportunity to try out and compare rafts.

Although there are many features to consider, it is important to make sure the raft on your boat meets the applicable regulatory requirements, and is made by a company that will stand by its product. In North Pacific waters, a raft should have an inflatable floor to insulate you from the cold water, and a canopy to protect you from the elements. If you are required to have a U.S. Coast Guard-approved raft, make sure yours meets that minimum standard before purchasing it.

Carefully choose the equipment you want packed in your raft; the coastal pack may not be enough for your situation, even if it

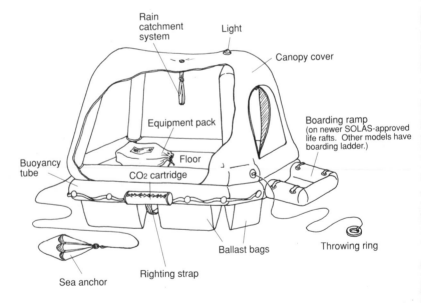

*Make sure life raft and survival suits are readily accessible,
and all crew know how to use them. (AMSEA photo)*

meets the requirements where you fish. The Sea Survival chapter has additional information on life raft equipment packs.

Mount the raft where it will be accessible in an emergency, yet protected from damage. Apply a non-skid surface to the deck near the raft's cradle, and make sure the raft can float free if the vessel capsizes. Do not place the canister near exhaust stacks; their heat and gases can damage the canister's rubber gasket and cause it to lose its water-tight integrity.

Once you have positioned the raft, secure its painter to the vessel. The painter is attached to the CO_2 inflator on the raft and must be pulled to inflate the raft. The painter has a weak link that is designed to break and allow the inflated raft to break free if the vessel sinks. Don't lash the raft to the vessel with extra line. You might not be able to cut it loose when the vessel is sinking. Avoid storing gear in or lashing gear onto the raft. You cannot afford to have anything interfere with the raft's deployment when you need it.

To help ensure that your raft will work when you need it, have it serviced once a year by an authorized agent. Manufacturers provide certificates to authorized repacking and service stations. Ask the raft repacker to show you the certificate from your raft's manufacturer.

An annual inspection is advantageous for several reasons. Out-dated flares and perishable items will be replaced, and wet rafts can be dried out. (In wet climates, water has a tendency to work its way into raft canisters.) Regular inspections can help extend the life of your raft.

Servicing also provides an excellent opportunity to see your raft—or one of the same design—inflated. Most servicing facilities welcome the opportunity to show off their product. Take advantage of it.

Information on launching rafts can be found in the Sea Survival chapter.

Immersion (Survival) Suits

Immersion suits have helped save hundreds of lives because they have excellent hypothermia protection (insulation) and flota-tion. To be most effective in an emergency, your suit must be main-tained, stored where it will stay dry, and placed where you can get to it in an emergency (**not** under your bunk). In addition, you must know how to put it on in a hurry. To be able to do this quickly in an emergency—especially in the dark—takes practice.

Wearing a survival suit will greatly increase your likelihood of survival if you have to abandon ship.

Consult the Personal Flotation Devices chapter for detailed infor-mation on selecting, donning, and maintaining immersion suits.

Emergency Signals

Thorough preparation also includes purchasing and maintaining appropriate emergency signals, and making sure all crew members know how to operate them. There are many emergency signals available to fishermen including: Emergency Position Indicating Ra-dio Beacons (EPIRBs), a "Mayday" given over your VHF or SSB ra-dio, flares, dye marker, signal mirrors, flags, strobe lights, reflective tape, and improvised signals. They all work by attracting a rescuer's attention and conveying your need for help.

It's a good idea to have a variety of signals to use in different circumstances because some are more effective at night, others dur-ing the day. Signals can be mounted on the vessel, stored on deck in a waterproof container, packed in your life raft, and kept in your survival suit pocket, your personal survival kit, and on your person.

The type and quantity of signals will be determined by your personal preference, the situation, Coast Guard requirements, and cost.

EPIRBs

Want a signal that's not affected by weather, darkness, or daylight, that constantly signals for help once it is turned on, and whose signal can be detected over great distances and can lead rescuers directly to you? If so, you need an EPIRB.

Experience has shown that the use of EPIRBs usually results in rescuers being on scene in less time than it takes when EPIRBs are not activated. Time is of the essence in an emergency.

To ensure that your EPIRB will work effectively:

- Install it according to the manufacturer's recommendations. Improper mounting can render it ineffective. Float-free types must have **no** obstructions overhead that would prevent their release.

- Test it at least once a month by following the manufacturer's instructions, and record the test date and results in a log book. The 406 mHz EPIRBs can be tested any time. All other EPIRBs should be tested for only 1 to 2 seconds during the first 5 minutes of any hour.

- Be sure the on/off switch is in the proper position whenever you leave port! For most 406 mHz EPIRBs this will be the "armed" or "automatic" setting. Your EPIRB's instructions will indicate the correct position.

- Replace the battery before it becomes outdated.

- For Class A EPIRBs: Make sure you put the plastic foam spacer (if there is one) on the correct side of the battery. If you don't, floating EPIRBs won't float upright, and antennas won't transmit because they'll be under water.

Register your 406 mHz EPIRB by sending in the registration card that comes with it or by contacting:

NOAA/DSD
E/SP3
4700 Silver Hill Rd., Stop 9909
Washington, DC 20233-9909
(301) 457-5678 or (888) 212-7283 (SAVE)

Upon registration NOAA will send you a dated decal. This registration system aids search and rescue personnel in finding vessels in distress and helps silence false alarms.

Non-Radio Emergency Signals

Non-radio signals can also be very effective. The strategic use of flares has helped rescuers locate many survivors, but flares can be dangerous if handled improperly. Outdated flares have a very high failure rate, so don't let them expire or they may not work when you need them. Become familiar with how your flares work. Some cannot be operated easily when you are wearing a survival suit.

Other emergency signals transmit or reflect light to communicate your need for help. Signal mirrors reflect a strong beam of light on sunny days, but can also work (although at a reduced intensity) in hazy weather conditions. Strobe lights are good at night—as long as the battery works and you are not trying to signal for help in a forest of longline gear that has a strobe on each pole. Don't underestimate the value of reflective tape. Some survivors have been rescued when a searcher's light reflected off the tape on their immersion suit or life raft.

An orange flag with a black square and black circle in the middle is an internationally recognized distress signal. Regardless of the signals you choose, they can't help you if you don't have them with you.

Both the Sea Survival and Shore Survival chapters have more information on signals.

Personal Survival Equipment

Will your clothes help keep you warm and afloat if you fall overboard or need to abandon ship? Although you may not think of it as such, your clothing and the items in your pockets are part of your survival equipment.

Clothes—like any shelter that helps retain body heat—keep you warm by trapping air. Fabrics vary in their ability to trap air and thus keep you warm, especially when they are wet.

Cotton is comfortable to work in, and can keep you warm—as long as it stays dry. However, cotton readily absorbs water and other liquids, and as soon as it gets wet it loses most of its ability to trap air—and keep you warm. Wet cotton clothing can quickly cool you, plus it takes a substantial time to dry.

Although they are not perfect fabrics, wool and polypropylene clothing might be a better choice than cotton because they can help keep you warm even when they are wet. Polypropylene dries much faster than either wool or cotton, but it is very flammable and leaves molten globs of fabric on your skin when it burns. Wool is heavy when wet, but it is not very flammable. Carefully consider your options. Proper clothing can make a significant difference in a survival situation.

Do you wear a hat or keep one in your pocket? Your body loses 50 percent of its heat through your head. A wool or polypropylene hat will help you retain body heat—even when it is wet.

Personal Survival Kits

Do you regularly carry items in your pockets that can help you survive if you fall overboard or have to abandon ship? If not, put some in your pockets or make up a personal survival kit in a waterproof container that's small enough to fit in your pocket or put in your survival suit.

When deciding what items to choose for your kit or pockets, concentrate on shelter, signals, and personal health considerations, and tailor the contents to fit your situation. Consider including:

- Shelter helpers such as nylon cord, wire, duct tape, plastic bags, space blankets, etc.

- Signals such as a mirror, whistle, surveyor's tape, brightly colored bandanas, small flares, fire starters, etc.

- Personal health considerations such as prescription medication, tampons, bug repellent, contact lens cleaner and holder, food, etc.

- Fire starters such as waterproof matches, a lighter, candles, magnesium fire starter, steel wool, and synthetic fire starting gel or sticks.

- Other items such as a good pocket or sheath knife, fish hooks and line, lures, aluminum foil, etc.

Whatever you choose, be realistic. It won't do you any good if you don't have it with you.

Training and Drills

Do you know how to operate the survival equipment on board your vessel? Studies indicate that training is more important than any other factor in determining whether an individual reacts positively in an emergency. Drills allow you to check your equipment, reduce reaction time and mistakes, and help diminish fear and panic.

Both scheduled and spontaneous drills can help hone a crew's reaction to an emergency. Drills can be fun, but they should not jeopardize anyone's life.

Regardless of the size of your vessel, all crew members should know how to:

- Abandon vessel.

- Fight a fire in different locations.

- Recover a person from the water.

- Minimize the effects of flooding.

- Launch and recover survival craft.

- Put on immersion suits and PFDs.

- Put on a firefighter's outfit and self-contained breathing apparatus (SCBA) if so equipped.

- Make radio distress calls.

- Use visual distress signals.

- Activate the general alarm.
- Report inoperative alarm and fire detection systems.

See the Man Overboard; Sea Survival; Fire Fighting; and Orientation, Emergency Instructions, and Drills chapters for information on responding to those emergencies. Basic navigation, seamanship, first aid, and CPR training are also important. Consult the resources section for more information on agencies that offer training.

The Will to Live

You can't buy, trade, sell, or touch it, but it's real, it helps save lives, and **you** can have it. It is the will to live—one of the most important aspects of surviving any emergency, any place, any time.

How you think and act can be as important as the equipment you have. Adopting the will to live means not giving up when the situation seems overwhelming or hopeless. It means deciding you **will** survive, despite the circumstances.

Preparation

Preparing for an emergency starts with the recognition that an accident is possible. Solid preparation will not only give you peace of mind, it also helps save lives, and can keep you out of the courtroom.

Reading the Weather

When you decided to fish in the North Pacific, did you know that it has some of the world's worst weather? Blame it on those clashing air masses above you.

Over the North Pacific, cold, relatively dry polar air sweeps south and collides with the warm, moist air over the Japanese current. In winter these air masses come together near the Aleutians and form great storms that move into the Gulf of Alaska. During the summer months the air masses meet nearer the Bering Strait, allowing the Pacific high pressure systems—and their accompanying good weather—to move into the gulf. Summer storms in the gulf are possible, but they are less intense than winter storms.

Do you know how to get the most out of National Weather Service broadcasts? Can you read weather maps? Do you regularly make and interpret your own weather observations? These can all help you make wiser fishing decisions. Understanding weather terminology is the first step.

Weather Broadcasts and Maps

High and Low Pressure

The air in the earth's atmosphere constantly moves to equalize temperature differences between the earth's air, water, and land masses. As these bodies of warm and cold air circulate, they build up areas of low and high pressure—represented on weather maps by the letters L and H. Low pressure systems usually bring poor or worsening weather, while high pressure systems generally produce good or improving weather.

On weather maps, high and low pressure systems have lines drawn around them. These lines—called isobars—connect places of equal barometric pressure. Isobars not only show the shape of a weather system, but also give some indication of surface wind strength. Closer isobars mean stronger winds because there is a greater difference in air pressure in a shorter distance.

In the Northern Hemisphere, detailed maps also show that air flows clockwise around high pressure systems, and counterclockwise around low pressure systems. This air flow is caused by the earth's rotation and is reversed in the Southern Hemisphere.

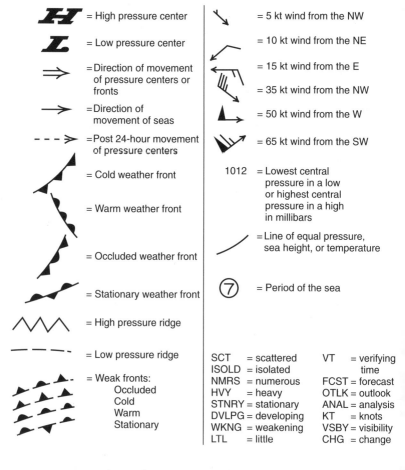

National Weather Service symbols and contractions.

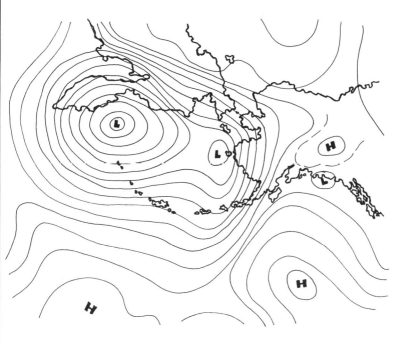

*High and low pressure systems over eastern
Asia, the Gulf of Alaska, and Alaska.*

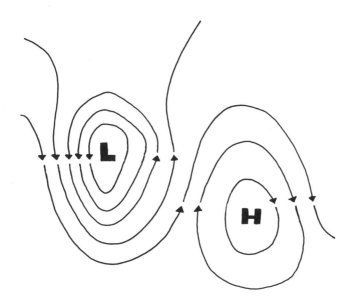

Air flow in the Northern Hemisphere.

Fronts

Cold and warm air masses do not tend to mix, so they usually have a boundary—called a front—between them. Fronts are important to track because that is where low pressure systems develop, with their accompanying deteriorating weather.

Meteorologists use symbols to distinguish the four kinds of fronts: cold, warm, stationary, and occluded. Cold air moving toward warm air is called a **cold front**, and is represented by a line with triangles on its leading edge. **Warm fronts**, depicted by a line with semicircles on its leading edge, occur when warm air moves toward cold air. When a cold front overtakes a warm front, the result is an **occluded front**, and is shown by a line with both triangles and semicircles on its leading edge. If a front becomes **stationary**, the semicircles and triangles are drawn on opposite sides of the line.

Storms

Although land formations can alter a weather system's course, the systems generally travel from west to east in the Northern Hemisphere. This, too, is caused by the earth's rotation.

The National Weather Service (NWS) issues three categories of wind warnings: small craft advisories, gale warnings, and storm warnings.

- A **small craft advisory** is a prediction for sustained winds (more than two hours) from 18 to 33 knots.
- **Gale warnings** forecast winds from 34 to 47 knots.
- **Storm warnings** predict winds of 48 knots and above.

The wind forecast also includes a prediction on wind direction.

The NWS issues sea height forecasts for coastal areas—for average sea height conditions to be encountered in open coastal waters unless otherwise indicated. The forecasted sea height is the average height of the highest one-third of the expected waves—although waves can occasionally combine and peak out at twice the forecast value. Sea height values do not take into account areas of normally higher or steeper seas found near bars, shoals, or restricted entrances into sounds or inlets.

Listening to Forecasts

Carefully listen to weather broadcasts at least two or three times daily and use either an actual or mental map of your area to picture where the weather fronts and systems are located and how they are moving in relation to you. Write down the pertinent information; memories are weak.

Marine weather broadcasts often include a synopsis of weather systems in the Gulf of Alaska and a 12 to 24 hour outlook. Sometimes a three to five day outlook will be included.

For a listing of NWS broadcast times and frequencies, and weather radiofacsimile (weather fax) transmission schedule, get a copy of the *Marine Weather Services Chart for Alaskan Waters* from your harbor master or the National Weather Service, 222 W. 7th Ave, Box 23, Anchorage, Alaska 99513, phone 271-5105.

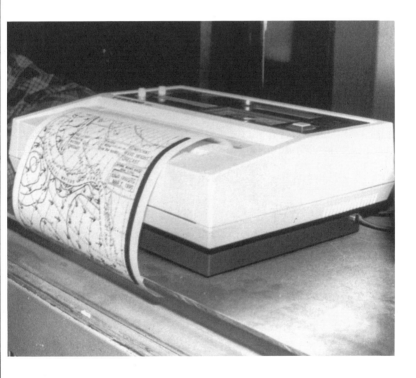

Weather faxes provide maps locating fronts and weather systems. Contact the National Weather Service for transmission schedule. (K. Byers photo)

Using Local Observations to Supplement Forecasts

Combine official forecasts with your own weather observations to avoid as much heavy weather as possible.

Barometric Pressure

As a General Rule:

- Falling barometric pressure means poor or worsening weather. The faster the fall, the stronger the storm.

- When a storm is upon you, the worst is over when the barometer starts to rise and the wind starts to switch from the south to the west, bringing windy, clearing weather. The wind direction can, of course, be modified by offshore winds out of bays, rivers, and passes.

- Rising pressure usually indicates clearing or good weather. The faster the rise, the sooner the clearing.

- When the barometer has been reading exceptionally high for a few days, weather changes occur slowly.

- Tides are somewhat affected by air pressure. The air in a high pressure system is **heavier** than in a low pressure, so it pushes down on the ocean more than a low does. Consequently, when the weather is dominated by a very high pressure system, the water level won't fluctuate as much, and high tides are often **lower** than predicted.

During a strong low pressure system, the air doesn't exert as much weight on the ocean, so the high tides can rise more. This, combined with the driving winds associated with intense low pressure systems, often creates damaging storm tides.

Reading the Barometer

Although a barometer does nothing more than measure air pressure, it can be a valuable tool, especially if you read it regularly, record the readings, and remember that air pressure is only one part of the total weather picture.

Although some barometers measure air pressure in inches, the National Weather Service measures it in millibars. See the millibar conversion chart in the NWS's *Marine Weather Services Chart.*

To get the most out of your barometer:

- Read it at least twice a day (morning and night are good times). Gently tap the barometer's face, and watch which way the needle moves. The tapping releases the stored up friction and usually makes the pointer jump slightly up or down, although the needle will move by itself with larger changes of air pressure.

- Record the reading, the time, and how much the pressure has changed since the last reading. This can go in your ship's log as a permanent record.

- After you have taken your reading, set the moveable needle on top of the pressure needle so you have a departure point for your next reading. This will tell you whether air pressure is rising or falling, and how rapidly it is changing.

Clouds

Observation of cloud types and movement is an important component of making your own weather predictions. Although this chapter does not detail the various types of clouds, some general observations are:

- Thickening and lowering clouds signal the approach of wet weather.

- High, thin clouds can be an early sign of approaching bad weather.

- Fair weather will generally continue when cloud bases increase in height along mountains.

Fog

- **Radiation fog** occurs in near-calm, clear weather when the earth loses heat into the night air, cooling and condensing the air above it. Depending on the wind, radiation fog can be anywhere from two to several hundred feet thick. It begins to evaporate shortly after sunrise, with the lower layers going first, but is slow to clear over water.

- When warm, moist air blows over cooler surfaces such as land or coastal water, the result is **advection fog**. This type of fog may develop day or night, winter or summer over the ocean and is the most likely type to be encountered at sea. Unfortunately, advection fog does not dissipate easily, and usually doesn't clear unless the wind shifts or increases dramatically in speed. Sunshine has no effect on advection fog over water.

- **Steam fog**, also known as **arctic sea smoke**, develops when air less than 10°F blows over warmer water. This type of fog is likely to occur where cold river water flows into the ocean or on inland bodies of water in the fall.

- If fog persists after 1 or 2 p.m. and no major weather changes have taken place, the fog will remain and probably become thicker.

General Weather Observations

- When the wind shifts to the west, the weather will generally clear.

- The weather will generally worsen when the wind shifts to the south or east.

- Fair weather will usually continue when there is heavy dew or frost at night, the moon shines brightly, and the wind is light.

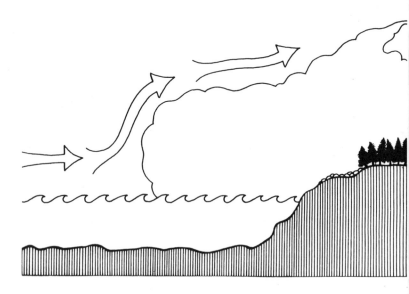

Development of advection fog.

The latter conditions are also often associated with a falling temperature.

- The temperature will usually fall when the wind shifts into or continues to blow northerly or northwesterly, or when the barometer rises steadily in the winter. Conversely, the temperature will rise when the wind shifts from the west or northwest to the south.

- To locate the center of the storm, use Buys Ballot's law of wind and pressure. In the Northern Hemisphere, face the true wind (not the apparent wind caused by the combination of the true wind and the wind caused by the vessel's own forward motion) and stretch out your arms. According to the Buys Ballot law, the center of the low pressure will be to your right and somewhat behind you. The center of the high pressure will be to your left and somewhat in front of you. This method can help you track the path of a weather system.

- In coastal waters, weather is modified by many local factors including mountain ranges, islands, glaciers, and coastal currents. Learn your local weather indicators.

- Make weather observations often, noting both the current weather and any changes that have occurred.

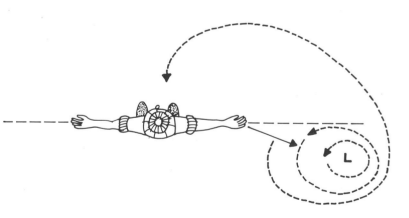

Locate the center of a storm by using Buys Ballot's law.
This law was formulated in 1857 by the Dutch meteorologist
Buys Ballot, and is also known as the Baric Wind law.

Icing

Vessel icing is potentially so dangerous that you should not rely solely on NWS icing warnings. Instead, you must be able to recognize conditions that cause icing and know both how to avoid icing, and what to do if it occurs.

Cold air, cold water, and wind create a deadly combination for vessel icing. Although icing can be caused by freezing fog, rain, or wet snow, the most common—and dangerous—source is from sea spray. You don't need to be underway to suffer sea spray icing. Boats at anchor are also at risk if they are in an area subject to strong, cold, offshore winds, especially off bays or near canyons or river mouths.

Nomogram charts that help predict ice accumulation are an invaluable aid and should be posted in the wheelhouse when icing is possible. Nomograms show that significant icing begins when the air temperature is less than 28°F (the temperature at which saltwater

*If your vessel ices heavily and you are in a
dangerous situation, call the Coast Guard and request
an escort to safe haven. (J. Dzugan photo)*

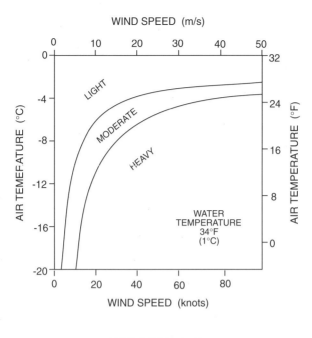

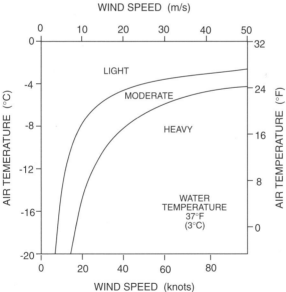

Nomograms of superstructure icing conditions due to spray, showing icing conditions for vessels heading into or abeam of the wind.

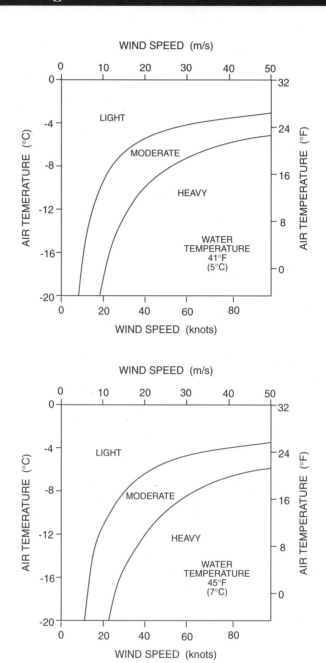

Light icing—less than 0.3 in/hr (0.7 cm/hr). Moderate icing—0.3–0.8 in/hr (0.7–2.0 cm/hr). Heavy icing—greater than 0.8 in/hr (2 cm/hr).

freezes), water temperature is less than 46°F, and winds are greater than 25 knots. The nomograms shown are based on recent evaluations, and predict icing rates up to three times greater than nomograms previously published by the National Oceanic and Atmospheric Administration. *Pacific Marine Environmental Laboratory Data Report No. 14* and *J. Climate & Appl. Meteorol.* 25(12):1793-1806.

Icing is dangerous because:

- It increases the vessel's weight. This both increases draft and reduces freeboard.
- The accumulation of ice on the vessel's superstructure and rigging raises the vessel's center of gravity and reduces stability.
- It can cause the vessel to list if the accumulation is uneven.
- It alters the vessel's trim.
- Speed and maneuverability are both reduced.
- The wind has more of an effect on the increased surface area of the rigging.

Although icing has a different impact on every vessel, vessels that are smaller, are marginally stable, take more water and spray on deck, have large deckhouses, or have heavy topside equipment or fishing gear are at a higher risk of capsizing due to ice buildup. Icing is most severe when you are abeam or heading into the wind.

Figuring Ice Weight on a Vessel

If you can estimate the surface area of the iced portion of your vessel, you can easily calculate the weight of the accumulated ice.

One cubic foot of saltwater spray ice weighs 60 pounds, which means that a 12" x 12" x 1" slab weighs 5 pounds.

To figure out how much the ice on your vessel weighs, you need to know both the surface area affected (area), and the speed at which the ice is building (thickness). If you are in conditions causing ice to accumulate at the rate of one inch per hour, how much would the ice weigh on a 10' x 20' area on your vessel's house after 1 hour? After 5 hours? The following equation will tell you:

Weight = area (in square feet) x thickness (in inches) x 5 pounds

Just plug your numbers into this equation:
The area (for this example) = 10' x 20' = 200 square feet
The thickness (for this example) = 1" after 1 hour, or 5" after 5 hours

This means that after 1 hour, the weight of the accumulated ice

on this 10'x20' portion of the house =
200 (area) x 1" (thickness) x 5 pounds = 1000 pounds
    The weight after 5 hours = 200 x 5" x 5 pounds = 5000 pounds
    The actual weight on this vessel will be greater in both cases because we only took into account a portion of the vessel. This extra weight will have a tremendous effect on how the vessel responds to handling and the sea conditions.

If Icing Is Anticipated

When the air temperature drops, consult nomograms to anticipate and avoid icing. When icing is possible, your main goal is to reduce your risk. Several possible actions exist:

- Run to a protected port.

- Head offshore to warmer waters.

- Slow your speed and alter your course to travel with, rather than into, the wind and seas.

When Icing Cannot Be Avoided

- Establish and maintain a radio communications schedule.

- Ready ice removal equipment (mallets and baseball bats) and rig lifelines on deck.

- Stop fishing. Store and secure fishing gear below decks or keep it secured and **covered** on deck—especially crab pot webbing and coils of line, which are notorious for collecting ice. If the situation warrants, your best course of action with crab pots may be to throw them overboard. Deck equipment should also be covered.

- Lower and secure the cargo boom.

- Prepare emergency and lifesaving equipment.

- Make sure scuppers and discharge pipes are clear.

- Batten down watertight doors, manholes, and hatches.

- Check deck lights to make sure they work. Icing will worsen at night if the temperature drops.

Combatting Icing

- If you cannot safely return to port, reduce your speed and alter your course so you are traveling downwind.

- Crew members working on deck should wear winter clothes and a Personal Flotation Device, and should be snapped onto a lifeline. They should be especially alert to the danger of frostbite.

- Remove ice first from life rafts, aerial wires, navigation lights, scuppers and discharge pipes, stays, shrouds, masts, rigging, superstructure, deckhouse hatch, anchor hoist, and net ports.

- Check the bilge regularly.

- If ice keeps building up in spite of your de-icing efforts and your roll gets lazy and sluggish, you've got too much ice on board. If your vessel ices heavily and you are concerned about your safety, call the Coast Guard and request an escort to safe haven. Have the crew ready to abandon ship. Sound drastic? Boats roll over very quickly, and you must be totally prepared with immersion (survival) suits on, life raft ready to launch, and no one below decks. This precaution is not pushing the panic button, only being sensible and ready.

Combine your observations with official forecasts to avoid as much heavy weather as possible. Seasoned fishermen know that heavy weather can turn a relatively uneventful trip into a life and death situation.

Handling Fishing Gear

He was new to crabbing and he made a bad mistake. Maybe it was caused by lack of sleep or not having eaten much all day. Whatever the cause, the break in attention was long enough for his left foot to catch in the coil of line as the pot went overboard, taking him to the bottom.

Seasoned fishermen know how dangerous their world is, but a green crew member or someone who's moving into a different fishery may not recognize the new hazards. Understanding and staying alert to dangers can help a fisherman keep his fingers, his gear, and his life.

General Points for All Fisheries

- Learn the ropes from experienced crew members, on both your own and other boats.
- Stand clear of working lines. Getting tangled up can give you a quick trip overboard or a ride to the hospital with a broken leg or mangled foot.
- Take time to clean the slime and fish waste off the deck.
- Stay out of the line of pull of cables.
- Keep a sharp knife handy to cut the web, line, or groundline in an emergency. This way, if you are snagged or caught as the gear is laid out, you will have a chance to cut yourself free before you go overboard.
- Wear a Personal Flotation Device (PFD) and non-skid boots while working on deck.

- Assume that decks will be slippery and move accordingly.

- Remember that drug and alcohol use, monotonous and repetitive work, and fatigue increase your risk of having an accident.

- Keep your mind on your work and try to anticipate what will happen next. It's easy to get hurt quickly.

- Watch out for the poisonous spike on the dorsal fin of ratfish and spiny dogfish. If you do get stuck, treat the wound immediately according to instructions in the First Aid chapter.

- Avoid sliding your hands along cables. Broken wires in the cables can tear away gloves and flesh.

- Don't stand or walk under suspended loads or pass a suspended load over someone.

- Protect yourself from jellyfish stings by wearing raingear, gloves, a hat, and goggles or glasses, and by washing your hands before you touch your skin. Some fishermen smear petroleum jelly on their faces to keep from being stung.

Stay clear of the pull cables.

Keep a knife handy when working on deck.

- Sprains and strains, especially of the back, continue to be the most common injury among Alaska fishermen for all gear types. Most of these injuries can be prevented by using proper lifting and moving techniques. The four most important ways to save your back are:

1) Use your brain first, and get help if you need it.

2) Before you pick up anything, get as close to it as possible, approach it squarely and bend your legs.

Watch out for fish spines.

3) When you stand up, keep your back as straight and square to the object as possible and let your legs do the work. One of the best ways to **hurt** yourself is to be twisted and bent at the spine when picking things up.

4) If you do a lot of forward bending, occasionally stop and lean backwards to stretch those muscles.

• Cuts, scrapes, and punctures are the second leading cause of injury to Alaska fishermen. Many of these could be prevented by wearing gloves and watching body placement. Longliners and

Lift properly to avoid back injuries.

trollers are more apt than other fishermen to have these injuries become infected, and most infections can be prevented by good hygiene. See the First Aid chapter for details on preventing infections.

- Bruises continue to be the third leading injury for seiners, gill-netters, and trawlers. Be aware of your body placement and stay out of the way of full nets to help prevent this sometimes crushing problem.

Crabbing

- Make sure pots are properly secured to the pot launchers.
- Take care when lifting pots out of the water or when moving them about, especially in rough seas. Swinging pots can kill people, so stand clear of them. Drop swinging pots to the deck or let them hit the outside of the bulwarks.
- Unload pots only on deck.
- When underway, secure pots to prevent them from shifting by tying each pot to other pots in at least two or three places.
- Know your vessel's stability limits. All boats should have a stability report on board that describes safe loading for given conditions and the proper loading pattern. You are always more stable with less weight above the waterline than with more.

Gillnetting

- When setting nets, watch for backlashes from the usually free-wheeling reel.
- Connect the power net reel's clutch control to a foot pedal that must be stepped on to operate the reel. Then, if someone is caught in the net, the reel will stop as he is lifted off his feet.
- Gillnets snag easily, so watch what you wear and avoid exposed buttons, buckles, or boot tabs.
- Stay off the rocks by using a fathometer with an alarm when you're drift net fishing.

Longlining

- Stay clear of flying hooks.

- When you're clearing gangions, wear wristers and raingear for protection against spinning hooks.

- Broken hooks can fly at you when you're working the roller, and spinning hooks can hit your face and cost you an eye. Wear goggles or a face shield to protect your eyes.

- Never grab the running gear outside the roller. If you must work the gear, stop the hydraulics first. Muscles tear and arms break when haulers grab line or gear outside the roller while pulling or setting gear.

- Big, flopping halibut on deck can be a serious hazard. Watch their tails, and stun and bleed them as quickly as possible.

Seining

- Keep your feet clear when the net's going overboard to make sure you don't go over with it.

- Rolling a big load of fish aboard or pulling a seine off a snag can easily cause a boat to roll over. Don't be greedy or impatient.

- Be very careful with your hands, feet, and clothing around the deck winch. You can be wrapped up and seriously injured in a flash and have no chance of stopping the winch yourself.

- Seine rings falling over the power block can knock you out and split your skull. So can a falling block. Either wear a hard hat or be aware of where you are standing.

- Don't let your clothing or body—especially arms and legs—get fouled in the net as gear goes through the block, or you'll get an unwanted ride.

- Don't wear exposed buttons, buckles, or other fixtures that can hang up on the web.

Trawling

- Stand outside the pull of the cable leading from the block at the railing into the winch drum. If the cable breaks, its whipping end might hit you—and divide you in half.

- Don't try to lead a cable onto the winch drum with your hands or feet. If the winch is not equipped with a proper levelwind, use an iron rod or length of pipe to guide the cable onto the drum. Then, be sure to have good, solid footing, because a fall across the cable could cause you to be carried into the winch.

- Take care not to lay too many turns on the gypsy head, especially if the head is short. An extra turn may cause overlap, particularly if the lead from boom point to winch is less than perfect. If the line shows the smallest tendency to overlap, remove a turn from the gypsy head.

- Wear a hard hat or safety helmet when on deck.
- When the net is being towed, don't sit or lean on the trawl cables.
- Stay alert when trawl doors are coming aboard, opening, or moving. Don't get crushed between the door and vessel.
- Use a lifeline whenever you enter the stern ramp or walk the codend.
- Keep clear of the codend if you split the catch on deck. You could be badly injured if it hits you as it's rolling or swinging.
- Don't step on the tackle as it comes down on deck from the gypsy head. Throw the tackle rope away from and clear of your feet.
- Following seas and a stern ramp can mean a dangerous deck, so keep the safety gate closed whenever possible.

Trolling

- Have a non-skid surface on the floor of the cockpit.
- Make sure there is a suitable breaking strap on the cannonballs to prevent them from hanging up on the bottom.
- When putting the gear out, make sure the lure is tossed over before snapping on the spread.
- Keep a pair of wire cutters of sufficient size in the trolling cockpit so you can snip your gear loose if it gets hung up. Otherwise, the seas and current might cause you to capsize.
- Secure the handle of the gurdy brake when stopping to work with the gear. Never try to stop the line with your hands or take snaps off when the gurdy is still running. Fishermen lose fingers left and right due to this carelessness.
- When there's a big fish on the spread, unsnap it from the wire to a kill-line or line on the boat before hauling it in.
- When placing cannonballs in their holders, beware of them swinging about and watch your fingers.
- Rig a method to get back on board, such as a floating, trailing line tied up high, in case you fall overboard.

Think Ahead!

Advanced medical care is tough to get at sea. Get some first aid training and think about what you are doing. Practicing common sense can prevent costly delays, injuries, and loss of life—and can keep you fishing.

Avoiding Fatigue

Would you give up half of what you own for a pill that would allow you to get by with only one hour of sleep daily? This question, raised by Dr. Gregory Stock in *The Book of Questions*, may seem like the answer to the commercial fisherman's dream. Unfortunately, no such pill exists.

Lack of adequate sleep is as much a part of commercial fishing as rough seas, sore hands, and black coffee. However, over the past few decades, increased competition, short openings, and large debts have pushed fishermen to work around the clock.

Working crews with little or no sleep may appear economically attractive but it can lead to serious problems. Tired people are careless, less attentive, less capable of making quick decisions, and a liability to themselves and fellow crew members. Hydraulic valves get turned the wrong way, hands go through blocks, fingers get cut by bait knives, boats run aground, knots get tied wrong, and gear or lives are lost.

Adding to these problems is something called boater's hypnosis: a fatigue brought on by just four hours' exposure to the noise, vibration, sun, glare, wind, and motion that occurs while on the water. The resulting fatigue slows reaction time as much as being legally drunk.

Extensive research has been conducted on sleep, partial and total sleep loss, and the effects of sleep deprivation on human performance. While these studies have found no cure for sleep deprivation other than sleep, they have led to a much better understanding of the problem and solutions that can be adapted to the fishing industry.

Effects of Sleep Loss

We all have an internal biological clock that establishes our daily rhythm of physical and psychological actions. Our bodies tend to want to follow this routine even when our normal sleep-wake cycle is altered.

Our alertness and stamina normally reach their daily low between 2 a.m. and 6 a.m., when we are usually asleep. When we don't sleep during these hours, our work performance bottoms out, the effects of sleep loss are amplified, and a large number of accidents occur. It is difficult to predict how sleep loss will affect an individual because everyone reacts differently.

The majority of researchers agree that the most significant effects of sleep loss are psychological. Missing just one regular sleep cycle can cause a deterioration in your mood: We have a tendency to become more irritable, depressed, disoriented, and unable to concen-

trate. Sleep experts have found several links between sleep loss and performance that will interest most fishermen:

- **Interesting jobs are less likely to be affected by sleep loss than boring jobs.** Fishing in extremely heavy seas or with high catch rates might be interesting enough to ward off many of the negative effects of lack of sleep. Conversely, slow fishing on calm seas with little sleep can make for such boring conditions that performance can go downhill.

- **New and complex skills are often more seriously affected by lack of sleep than simple tasks that are second nature.** When you haven't had enough sleep, the more troublesome skills are those that are longer, more complex, newly learned, or not well practiced. A wise skipper should keep an eye on new, tired crew members.

- **As sleep loss increases, performance becomes more uneven.** The poor performance that results from lack of adequate sleep is more likely to be sporadic than continuous. For fishermen, this means that a tired crew member who is performing well—and appears to be okay—may lapse into unsafe behaviors with no warning.

- **Sleep loss slows reaction times.** Lack of sleep leads to slowed reactions, a dangerous situation when precision is needed to work with heavy equipment in rough seas.

- **Most people take 10 to 15 minutes to fully function after waking up.** The groggy muddiness of mind that most of us feel just after we wake up significantly affects our ability to work.

Solutions

One of the major conclusions from sleep loss research is that the only sure way to completely recover from sleep deprivation is—you guessed it—to sleep. Five to twelve hours of uninterrupted sleep allows most people to recover from even many days without sleep. You can take steps to minimize the effects of sleep loss.

Before an Opening

The 12 hours prior to a long period with no sleep—such as before an opening—should ideally be spent resting or sleeping to help minimize later problems. Prepare for an opening ahead of time instead of leaving all the details to the last minute. Then get some rest.

While Fishing

Because the effects of sleep loss are largely psychological, anything that increases psychological arousal, or enhances mood and motivation, improves job performance. This is especially true during the 2 a.m. to 6 a.m. lull.

One tried-and-true technique for reducing boredom, and thus increasing work productivity, is periodic job rotation. The ideal situation is to cross-train the crew for all deck jobs so they can swap back and forth. Interesting jobs are less likely to be affected by lack of sleep than tedious jobs.

Good communication between the skipper and crew also improves work performance. Crew members need immediate feedback—both positive and negative—on how they are performing, when their next break will be, etc.

Wake crew members at least 15 minutes before they are needed to stand watch or perform other activities that require them to think.

Ways of increasing arousal include moderate exercise (such as jumping jacks to increase circulation and oxygen intake), listening to high-spirited music, splashing cold water on the face, chewing flavored bubblegum, and drinking soda pop and hot drinks.

Brushing your teeth, washing up, or quickly shampooing your hair can improve your mood and help keep you awake. Stimulating the sense of smell with aftershave, soaps, and hand lotion can also have a refreshing effect. Good hygiene before hitting your bunk has the added advantage of helping you sleep better.

Humor can be extremely effective in improving mood, as can contests and safe games between crew members.

Practice Safety Procedures

Accidents often happen when crews are tired and performing poorly. Hold fire drills and abandon ship drills often enough so they become second nature, and the crews are not as likely to be affected by sleep loss.

Critical Tasks

When working without enough sleep, it is important to pay extra attention to critical tasks such as running the hydraulics, leading buoys through the block, and snapping gangions. All senses are needed during these potentially dangerous times.

Working Together

Military studies have shown that the central factor that determines the overall performance of a group is unit cohesion—how well everyone works together. This is also true for fishing crews. Important components of a tight crew are trust and confidence in the skipper, fellow crew members, oneself, and the vessel. The best skippers know that there are better ways to motivate a crew than fear and pain, and that tight crews should be better able to resist the effects of sleep loss because of the positive psychological environment they create.

Naps

Naps can be very beneficial, but to be most effective they should be taken during the 2 a.m. to 6 a.m. low. Otherwise, crew members might be too wired to fall asleep quickly even though they are very tired. This sort of insomnia can be caused by anxiety or fear of missing something.

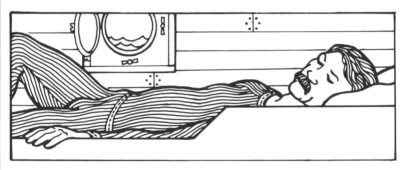

Naps help improve performance.

Eat Well

Bodies—like boats—need fuel to run. For people the fuel is food, and good food is premium. Eating well is especially critical when you are low on sleep.

A good breakfast is especially important. If you want to be irritable, moody, depressed, uncooperative, and slow-performing, skip breakfast and eat foods high in carbohydrates (this includes sugar!). Eating too many carbohydrates causes your energy level to rise for a while, but then fall very rapidly, often making you feel more tired, grouchy, and slower-thinking than before you ate.

If you want to feel better and have a sustained level of energy, eat meals that include protein, carbohydrates, and some fat. Protein is found in foods such as seafood, meats, nuts, milk and milk products, beans, rice, and soy products like tofu. Grains and cereals, bread, pasta, vegetables, beans, and many fruits are high in carbohydrates. Foods with high fat content include whole milk products, red meats, nuts, oils, fats, shortenings, and many desserts and breads.

Try eating sandwiches, nuts, or fruit for quick snacks instead of high fat, high sugar doughnuts, candy bars, cookies, or other junk foods. Your tired body will be thankful.

Gain energy by taking the time to eat.

Remember . . .

Until openings no longer require sleepless days on end, fishermen need to know how to reduce the problems associated with fishing without enough sleep. Understanding how sleep loss lowers performance, and taking steps to decrease these problems, can make the difference between getting to port tired or arriving with a seriously injured or dead crew member.

Personal Flotation Devices

It happens more than we like to admit. Fishermen end up in the water when they don't want to be there—some because they had to abandon ship, others because they were drunk, a surprising number because they were standing at the rail relieving themselves.

Whatever the cause, the result is the same—the water is cold and it robs heat from their bodies 25 times faster than air of the same temperature. This cold water can cause even expert swimmers to drown or to die from hypothermia if they aren't wearing a suitable Personal Flotation Device (PFD).

If you cannot guarantee you won't end up in the water, you must know how to prolong your in-water survival time. This is where PFDs play their part.

Rick Laws is one of hundreds of people who owe their lives to a PFD. He was a commercial fisherman on the F/V *Cloverleaf* out of Kodiak when it sank near Sutwik Island. Laws spent 27 harrowing hours in rough seas in his survival suit before being rescued by another fishing boat. He readily admits that he would not be alive today if he hadn't worn the suit.

Many fishermen feel that PFDs are uncomfortable to wear and difficult, if not impossible, to work in. Today that's not necessarily true. If you ordinarily wear a vest, rain gear, a jacket, or suspenders when you fish, you can be wearing a PFD. This chapter examines both USCG-approved (U.S. Coast Guard) and non-approved PFDs, their maintenance, and advantages and disadvantages for North Pacific fishermen.

Choosing a PFD

Consider the following factors when you evaluate your PFD options: buoyancy, hypothermia protection, fit, comfort, visibility, cost, features, and legal requirements for your vessel.

Keep in mind that the more pounds of buoyancy a PFD has, the higher out of the water you will float, thereby increasing your chances of survival—especially in rough seas. (Pounds of buoyancy has no relationship to how much a PFD weighs. It refers to the Archimedes principle: that an object is buoyed up by a force equal to the weight of the water it displaces.) The only way to determine if a PFD will float you is to wear it in the water—preferably a pool because our cold waters can make it dangerous to try out PFDs that have little or no hypothermia protection.

Hypothermia protection is highest when the five high heat loss areas—the head, neck, armpits, sides of the chest, and groin—can be kept warm and dry. Some PFDs provide good hypothermia pro-

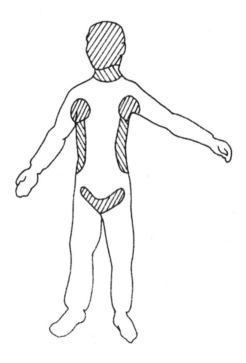

Five areas of high heat loss.

tection in calm waters, but are less effective in rough seas because body movement or the PFD design permits cold water to flush past the skin. This is one reason to make sure the PFD fits before you buy it. Try it on while wearing the clothes you normally wear on board.

Although you may not want to be seen in a brightly colored PFD, it will be more visible than a dark one and, like the use of reflective tape, will increase your likelihood of rescue.

PFDs are like parachutes—they only save your life if they are worn and they work!

Types of PFDs

Survival/Immersion Suits

Survival suits have helped save hundreds of lives, largely because they provide considerable hypothermia protection to the body's five high heat loss areas. A good immersion suit (the newer name for survival suits) that fits well can actually keep you dry. All USCG-approved immersion suits have a minimum of 22 pounds of buoyancy and are constructed so the wearer will float, even if the suit is full of water. Immersion suits provide the best hypothermia protection of any PFD currently on the market, but they will not turn an unconscious person face up in the water.

Although it is not practical to wear a PFD all the time, you may want to when work needs to be done in a dangerous situation such as breaking ice, or when preparations need to be made to abandon ship. If the suit does not have detachable or unzippable mitts, you can put it on your lower body and tie the arms around your waist.

The first suits manufactured had attached three-finger mitts but some now come with five-finger gloves, detachable mitts, or mitts that can be unzipped to free the hands. The skipper of the F/V *Cape Chacon* was the only person aboard whose suit had detachable mitts when his vessel sank in the Gulf of Alaska. With his free hands, he was able to tie and untie lines, help zip others into their suits, and fire smoke flares—tasks his crew members were unable to perform with their mittened hands.

Difficulty using mittened hands is a common problem in survival situations. When considering whether to purchase detachable or zippable mitts, keep in mind that even though they have some advantages, they may not be best for you. If the wrist seals do not fit tightly, water will be able to enter the suit. In addition, some people quickly lose function of their hands in cold water and would be better off with warmer hands in clumsy mitts.

All USCG-approved suits are required to have an attached air bladder or flotation ring that can be inflated by mouth. Make sure you leave the bladder or ring on the suit. Once inflated, it makes you even more buoyant and can help save your life. The placement, size, and shape of the bladders vary. Look for a bladder that helps

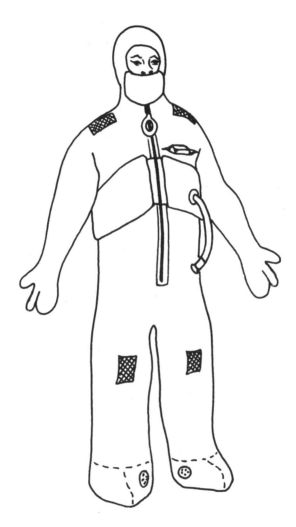

Immersion (survival) suit.

raise your head and shoulders well above the water level—an especially important feature in rough seas.

Know how your suit's air bladder inflates. If the hose's mouthpiece has a knurled ring, the ring must be screwed **away** from the mouthpiece or air will not be able to enter the bladder. Screwing it toward the mouthpiece prevents air from entering and escaping. Leave the knurled ring screwed away from the mouthpiece when storing the suit. A few of the newer model immersion suits use a CO_2 cartridge as a second way to inflate the air bladder.

Some of the tubes leading to the air bladders, especially the non-ribbed ones, have a tendency to kink. Other tubes are barely long enough to permit inflation. Try on your suit before you buy it.

Proper fit is essential for survival suits to work best. When the F/V *Wayward Wind* sank in the Gulf of Alaska, one survivor was in a suit too large for her. Water entered the suit through the space between her head and the top of the hood because she had settled down into the suit. This settling action caused the suit's face flap to push up over her nose and face, making it difficult to breathe and impossible for her to inflate the flotation ring.

If your fishing vessel has several different sizes of immersion suits on board, clearly mark them with tape or string so they can be easily distinguished in an emergency.

The only way to know for sure how your suit will perform is to

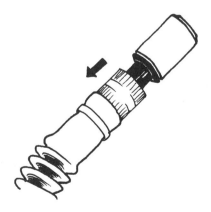

*Make sure the knurled ring is screwed away
from the mouthpiece of the immersion suit inflator,
so you can blow air into the suit's bladder.*

try it on with your normal work clothes and go in the water. Once you realize how well the suit keeps you afloat, you'll feel more comfortable in a survival situation. If you wear glasses, decide what to do with them when you are practicing. Don't wait until an emergency occurs to choose whether you will wear them or secure them in a shirt pocket.

Quick Donning Technique for Immersion Suits

The quick donning technique described below has proven effective for fishermen who need to put on their suits in a hurry in rough seas. You may develop a different quick, safe technique for your type of suit. Whatever method you use, **practice it before you need to do it in an emergency.**

Make sure you have on long underwear, pants, a shirt, and a jacket or sweater before donning the suit. Several layers of clothes underneath the suit will help keep you warm. Wool and polypropylene will keep you warmer than cotton if you get wet.

If you are likely to be wearing a PFD on deck, find out whether or not it will fit under your immersion suit. Some will, others will not.

Sit down to put on your immersion suit.

Designate an exterior location on board for donning immersion suits. Keep in mind that hand grabs may be useful. The area should be kept relatively dry and free of sharp objects that might damage the suit.

Get the suit out of the bag using a quick flick of the wrist. (Suits should **always** be stored unzipped.)

Sit down (it's important not to stand in rough seas) and work your legs into the suit. **Leave your boots on or put them in the suit with you**—you'll need them on shore. Your boots will slide into the legs of your suit with ease if you quickly slip plastic bags over your feet before putting them in the suit. These bags can be stored in the suit's hood for ready access.

While still sitting or kneeling on deck, place your weaker arm in the suit. If you are right-handed, this is usually your left arm. Then pull your hood on with your free hand. It's hard for some people to pull the hood on with a gloved hand, which is why the hood is put on now. If your suit has detachable mitts, it may be easier to put your hood on after both arms are in the suit.

Place your stronger arm in the suit last. Pull the zipper all the

Put your weaker arm first in the immersion suit.

way up, then secure the face flap over your face. (If you can't easily grab your zipper-pull with gloved hands, use non-rotting line to secure a piece of dowel or other grabbable item to the pull.) Help other crewmembers with hoods, zippers, and face flaps.

If you will be entering the water from a height, wait until you are in the water to blow up the flotation ring or bladder. Blowing it up and then jumping in can cause a neck or back injury, or could rip the bladder off the suit.

In-water Donning Technique for Immersion Suits

It is possible to get into an immersion suit in the water, but the cold temperature makes it both difficult and potentially dangerous. **If you are going to practice this method, do it in a pool.**

Try to keep your head dry—you lose 50 percent of your body's heat through your head.

Lay the suit face up in the water. Straddle it, lean back, and kick your feet in, pulling the suit up around you. Next, put in your weakest arm, pull on the hood, and put in your other arm. Then zip the suit up and secure the face flap.

You will have lots of water in the suit, but you cannot sink. As long as the water is not being flushed in and out, the suit should act like a big wet suit, and your body will warm the water. You will cool off some, but you will be warmer than if you were in the water without an immersion suit. Obviously, it's best to put your suit on **before** you go in the water.

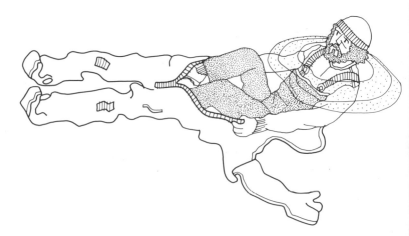

To put on an immersion suit in water,
straddle it and kick your feet in.

Type V PFDs—Coveralls

Also known as work or deck suits, coveralls are a Type V PFD. With a minimum of 22 pounds of buoyancy, some coveralls are approved to substitute for a Type III PFD when they are worn. Although the coveralls will not keep you dry if you end up in the water, they still provide fair hypothermia protection, especially if the waist and leg straps, and velcro around the wrists and ankles, are snug.

Most coveralls have an inflatable pillow that will help keep your head out of the water, but coveralls will not right an unconscious person in the water. Pillow inflation varies from suit to suit, so make sure you know how to work yours. If you have the type with a knurled ring, make sure the ring is screwed away from your mouth or you won't be able to blow in air.

Because coveralls can be difficult to put on in a hurry, it is best to wear them if they are your PFD. You can, however, be quite hot in them if you are very active. They are especially useful when you are traveling at planing speeds in an open boat.

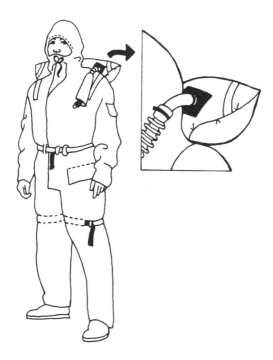

Many coveralls have an inflatable pillow.

Type V Hybrid PFDs

The newest USCG-approved PFD is the Type V Hybrid, which combines 7.5 pounds of inherent buoyancy with an air bladder that can be inflated with a CO_2 cartridge or by mouth. This may be the most comfortable USCG-approved PFD to wear while working because it contains the least amount of inherently buoyant material but still provides 22 pounds of buoyancy when fully inflated.

The Type V Hybrid is designed for people weighing more than 90 pounds and, like the coveralls, meets U.S. Coast Guard requirements only if it is worn. It provides some hypothermia protection, though not as much as coveralls, float coats, or immersion suits. Maintenance and care of the inflating devices are critical if this PFD is to perform as intended. Type V Hybrids are not being manufactured (in 1998), thus new ones are not available.

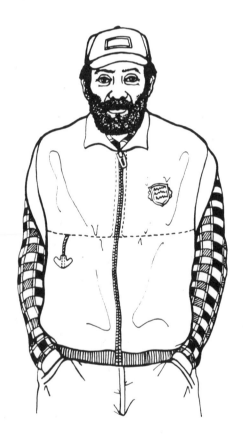

Type V Hybrid PFD.

Type IV PFDs—Throwable Devices

Throwable devices, such as life rings or cushions, are classified as Type IV PFDs and have 16.5 to 32 pounds of buoyancy. Although they offer no thermal protection, some allow you to get more of your body out of the water than with many other PFDs.

Keep life rings within easy reach to throw to an overboard crew member. The addition of a flagpole, coil of floating line, PFD light, and reflective tape at four points on both sides of the ring will make it easier to spot the life ring and to haul the person back on board.

Commercial fishing vessels less than 65 feet long are required to have 60 feet of attached line on their life rings. Vessels over 65 feet are required to have 90 feet of attached line.

Type IV PFD—Life ring.

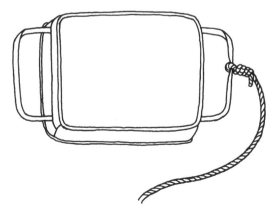

Type IV PFD—Cushion.

Type III PFDs—Flotation Aids

There are many different styles of Type III PFDs (Flotation Aids), all of which have a minimum of 15.5 pounds of buoyancy. Some fishermen use the vests while working on deck because they can be worn under or over rain gear and allow fairly good mobility. Unfortunately, they will not right an unconscious person and they offer less hypothermia protection than many other PFDs do.

Some vests tend to ride up when worn in the water, although adjusting the vest's shoulders and sides can partially eliminate the problem. A few models have a waist strap that helps secure the vest. Beware, however, of the potential danger of getting this PFD's strings or straps snagged or caught by hooks, nets, etc.

The float coat, with built-in insulating and buoyant foam around the trunk, is another Type III PFD. Some float coats have an attached hood, insulated arms, and a neoprene beaver tail (also called a dia-

*Type III PFD—Flotation
Aid—Vest.*

*Type III PFD—Flotation
Aid—Float coat
with beaver tail.*

per) to lessen heat loss from the groin area. When secured, the beaver tail also helps keep the coat from floating up around your neck.

The float coat's good hypothermia protection often makes it too warm to wear while working on deck, but it may be the answer for those fishing in skiffs in near-coastal areas or rivers.

Type II PFDs—Nearshore Buoyant Vests

Type II PFDs (Nearshore Buoyant Vests) have a minimum of 15.5 pounds of buoyancy and will turn about 40 percent of the people who wear them face up in the water. Type IIs offer little hypothermia protection and are awkward to wear in many work situations, which is why most commercial fishermen are unlikely to rely on a Type II for their PFD needs.

Type II PFD—Nearshore buoyant vest.

Type I PFDs—Offshore Life Jacket

With a minimum of 22 pounds of buoyancy, Type I PFDs (Offshore Life Jackets) contain the most inherent flotation for their size. They are designed to float most people (about 80 percent) face up in the water, a big advantage if you are unconscious or unable to right yourself. However, Type I PFDs (also sometimes called "Mae Wests") provide only minimal hypothermia protection, and many people consider them too bulky for work on deck. This PFD is reversible—a nice feature if you need to put it on quickly. It does not provide as much hypothermia protection as float coats, coveralls, and immersion suits.

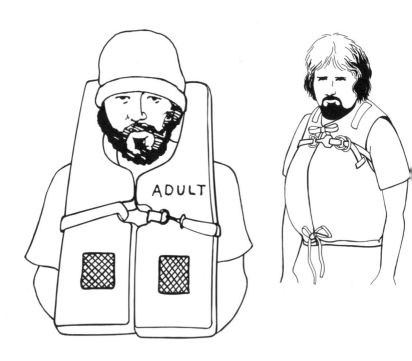

Type I PFD—Offshore life jackets.

PFDs Not Approved by the U.S. Coast Guard

There are several PFDs that do not meet USCG specifications but they may appeal to fishermen. These PFDs are very wearable, but some must be inflated to provide buoyancy.

The Stormy Seas™ jacket resembles an outdoor jacket and has an air bladder that can be inflated by a CO_2 cartridge or by mouth. Although the jacket has no inherent buoyancy, it is very comfortable to work in and will help keep you afloat for quick rescue.

Some manufacturers also sell a rain coat with an inflatable air bladder. Like any wearable PFD not secured to your legs or crotch, these rain coats tend to ride up when they are inflated.

Stormy Seas™ jacket.
(Not approved by USCG.)

Inflatable suspenders are another non-USCG approved PFD. Some models have three loops for a belt, while others are secured with a crotch strap. The suspenders can be inflated either by mouth or with a CO_2 cartridge. They offer no thermal protection but, like the jackets and rain coats with inflatable bladders, they can help a conscious person stay afloat for quick rescue.

SOSpenders™ uninflated.
(Not approved by USCG.)

SOSpenders™ inflated. (Not approved by USCG.)

PFD Maintenance

Maintain your PFD as if your life depends on it—because it does. Regular inspection and maintenance not only prolongs the PFD's life, but ensures that it will work when you need it. Avoid leaving PFDs where they will be exposed to fuel; this can cause some types to deteriorate.

Survival/Immersion Suit Maintenance

After the sinking of the F/V *Wayward Wind*, an investigation by the National Transportation Safety Board (NTSB) revealed that corroded zippers on the survival suits worn by three of the deceased crew members most likely contributed to their deaths. Because the suits were not fully zipped, cold water was able to flush in and out. In addition, inflation bladders were not attached to all of the suits and none of the suits had lights. The NTSB concluded that the crew members might very well have survived if the suits had been properly maintained.

Visually inspect your suit and bag for rips and tears once a month. Make sure the suit is dry. Storing these suits wet or damp can cause them to mold.

Shake your suit out of its bag every three months for a more thorough inspection:

- Examine it for rips and tears, and check the seams to make sure they are securely glued. Small leaks and tears can be repaired using a dive suit cement available at dive shops.

- Try the suit on, zip it up, secure the face flap, and blow up the air bladder. Does it still fit? Is the air bladder's tube securely attached to the bladder? Deflate the air bladder before storing the suit and make sure the knurled ring (if there is one) is screwed **away** from the mouthpiece so you can easily inflate the bladder in an emergency.

- If your suit has a CO_2 inflation system for the air bladder, check the CO_2 cartridge. Unused cartridges have no hole in their end, while used cartridges will have a hole. Test the inflation system once a year to make sure it works, and replace the spent cartridge immediately with an unused one of the **same size**. Using too large a cartridge can blow out seams in the bladder. Before installing a new cartridge, put the firing lever in the "up" position or the new cartridge will discharge when it is screwed in.

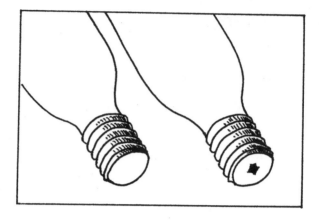

Unused (left) and used (right) CO$_2$ cartridges
for inflating PFDs and immersion suits.

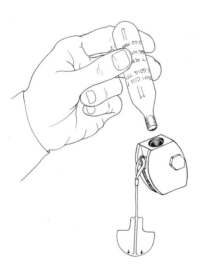

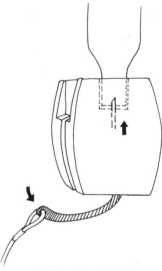

Make sure the inflator
mechanism lever
is up before installing
a new cartridge.

Pulling the inflator
mechanism lever down
pushes a pin into the
CO$_2$ cartridge.

- Do the PFD's whistle and light work? Replace malfunctioning PFD light batteries and batteries that will expire before your next inspection.

- Check the reflective tape. Is it securely attached, and does it reflect light? You should have 31 square inches of reflective tape on both front and back.

- Repair or replace anything that does not work. Make major repairs only with the manufacturer's approval, and repair and replace zippers only at authorized facilities. Non-approved alterations may jeopardize the suit's USCG approval.

- Wear the suit in the water at least once a year to check for leaks. When your suit has been used in salt water or a swimming pool, rinse it inside and out in fresh water. If it has been contaminated with petroleum products, wash it by hand in warm, soapy water (use dish or hand soap), and rinse it thoroughly.

- Turn the suit inside out, dry it completely in a shady, well-ventilated place, then turn it right side out to dry the outside.

- Relubricate the zipper according to the manufacturer's instructions. (Manufacturers differ in their recommendations.) Avoid petroleum-based greases or waxes; they can destroy the rubber on the zipper. Lubricate the inside of the teeth, too, and work the zipper several times, checking to make sure the zipper-pull is still securely attached.

- Leave the zipper **open**, and roll the suit up from bottom to top,

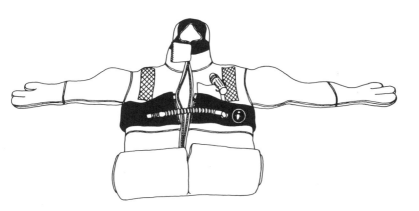

Roll your suit up with the zipper unzipped and air bladder deflated.

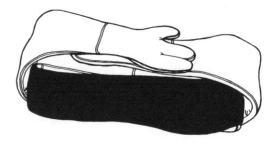

Avoid creases in the immersion suit's arms by laying the arms over the top of the rolled suit.

trying to avoid any folds. Lay the arms over the rolled-up suit, put the suit back in the bag, and write the inspection date on the bag.

- **Store the suit in a dry, accessible place where you can get to it quickly in an emergency.**

Maintenance of Other Kinds of PFDs

- Take care to maintain your PFD so it will keep you afloat when you need it.
- To prevent rot and mildew, thoroughly dry PFDs—both inside and out—before they are stored. If they have been soaked by salt water, rinse them in fresh water before drying to help prevent the zippers, zipper-pulls, snaps, and other metal parts from corroding. Follow the cleaning instructions on the label.
- At least once every three months—more often if you wear the PFD frequently—check it for rips, holes, corrosion, and decaying fabric. Although a small tear or hole will probably not destroy the garment's flotation, it should be repaired.
- Lightly lubricate metal zippers with a silicone lubricant designed for diving-gear zippers. Replace broken zippers, snaps, and fasteners.
- Check all straps to make sure they are still attached and in good condition.
- Try the PFD on. Does it still fit?
- Test the reflective tape to make sure it reflects light, and replace any defective tape. Be sure there are 31 square inches of tape on both front and back of each PFD.

- Many PFD manufacturers use the vegetable fiber kapok for the filling in Type I PFDs and seat cushions, sealing the kapok in plastic bags to keep it both dry and buoyant. If these kapok-filled bags get punctured or burst, the kapok can absorb water, and the PFD will lose flotation or may not float at all. Check the bags by gently squeezing the PFD, listening for air leaks. **If the PFDs leak, are waterlogged, or do not float, they should not be used**. Slice through and discard them to prevent others from using them.

- Check CO_2 cartridges and replace them as needed.

- Check life rings to make sure they float, and inspect the attached lines. Are they secured at all points? If faulty lines cannot be re-attached, destroy and discard the life ring.

- Be sure your vessel's name is clearly written on the ring.

- Test PFD lights to make sure they work, and replace defective parts or batteries that will expire before your next inspection.

Additional Equipment

Although PFDs have saved thousands of lives, keep in mind that reflective tape, PFD lights, EPIRBs, flares and other signals, and a personal survival kit attached to the PFD will improve your chances of survival and rescue. Some fishermen have been rescued because a searcher saw light reflecting off a four-inch strip of reflective tape!

Man Overboard!

Frank's nightmare had come true. One second he was on deck, the next he was in the water with his vessel heading out of sight in the fading daylight. Fortunately, only five or ten minutes passed before another crew member looked for and couldn't find him, then realized he must be overboard.

Frank was rescued, largely because the crew had practiced what to do in this situation. Once aboard, Frank was cold, wet, and embarrassed, but thankful to be alive. And glad he had been wearing a Personal Flotation Device (PFD).

There is no denying it. Fishermen do end up overboard. Between 1981 and 1985, 20 percent of all commercial fishing deaths in Alaska were due to falls overboard. You might never face this situation, but being prepared in case it does occur could save a life—maybe yours.

Before Anything Happens

- Hold man overboard drills with a crew member in a survival suit playing the part of the person in the water. The first drill can even be held in the harbor. You will need to devise a retrieval system; hauling someone on board is difficult and can result in injuries. Some forethought will allow you to come up with an effective system. Try a sling or horsecollar used with a winch and boom, a net with floats, a commercial man overboard retrieval device, or a hatch cover and winch.

- Make yourself warmer and more visible in the event you do fall overboard by wearing layers of wool or polypropylene clothing, and bright-colored rain gear or a PFD. Increase your chances of being rescued at night by putting **reflective tape** on your PFD, rain gear, hard hats, and anything else that might fall overboard. Reflective tape is a very cheap safety supply.

- Whenever you are on deck, wear a PFD with a whistle or other signaling device.

- Keep "one hand for the ship and one for yourself," especially when relieving yourself at the rail.

- Keep a throwable PFD—with a working light and reflective tape—handy to toss over as a marker in case someone does go overboard.

- Use a buddy system on deck after dark, especially in rough seas.

- Where practical, have bulwarks or safety rails of adequate height, and provide grab rails alongside or on top of the house.

- Use non-skid deck coatings, and keep decks clear of slime, oil, jellyfish, and kelp.

- Use safety lines when clearing ice.

- If you fish alone in a smaller vessel, consider using a kill switch that will shut off your outboard motor if you fall overboard. Or you may want to use a harness with a safety line that would keep you attached to the boat if you go overboard.

- Consider towing a skiff or knotted, floating line with a buoy on its end to provide a close target to grab or climb onto. If you use either of these methods, you **must** use **floating line** attached high up on the vessel, and be conscious of the possibility of the line getting fouled in the prop. Some fishermen believe in and use the skiff or floating line method, while others think they present more danger than they are worth. The choice is yours.

How Your Body Gains and Loses Heat

Heat Loss

- Radiation: When the air is cooler than your body, you lose heat through your skin.

- Respiration: You lose heat by exhaling air your body has warmed.

- Evaporation: Sweat evaporates your body's heat and moisture into the air.

- Convection: Heat radiated from your body is taken away by moving cold water and cold air.
- Conduction: Being in contact with cold surfaces takes your body heat away. Knowing how your body loses heat makes it easier to understand why **you lose heat 25 times faster in water** than in air of the same temperature.

Heat Gain

- External sources.
- Conversion of food to heat.
- Muscular activity: Note, however, that activity in cold water may cause you to lose more heat than you gain.

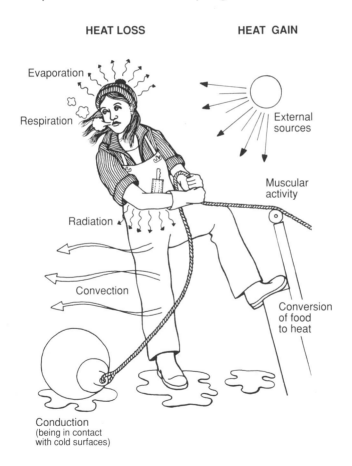

HEAT LOSS   **HEAT GAIN**

Evaporation

Respiration

External sources

Muscular activity

Radiation

Convection

Conversion of food to heat

Conduction
(being in contact
with cold surfaces)

When a Person Is Overboard

If You Are Overboard

1) Immediately yell for help or blow a whistle to attract attention.

2) Assume the Heat Escape Lessening Posture (H.E.L.P.) to improve your chance of survival.

3) Keep your clothes and boots on. They will keep you warm and will **not** pull you down.

4) Hold on to available floating objects that will help increase your buoyancy and make you more visible.

5) Stay as still as you can. (Rough seas can make this difficult.) Movement cools you off quicker because it uses energy and brings more cold water in contact with your body.

Your job is to stay afloat, conserve heat, and signal for help.

H.E.L.P. position.

If You Are Aboard the Vessel

1) As soon as someone is known to be in the water, record the location on loran, and throw something overboard to mark his position. A PFD works well because it also offers the person in the water more flotation. Be sure to turn on the PFD light. The position can also be marked by newspaper, buoys, a halibut pole with a radar reflector, or throwable smoke. Radar reflectors are especially helpful in foggy conditions.

2) Post a lookout whose sole job is to keep the person in the water in sight, and to point at him. A lookout is **critical** when you consider that one-foot seas make it difficult to see a crew member's head bobbing in the waves. Sound the alarm.

3) Turn the vessel around so the stern swings **away** from the side the person is on. Before you can approach the person you must find him. If he is not in sight, retrace your path and, if it has been more than a few minutes since he was last seen, notify the Coast Guard. They may be able to assist in searching and will notify other vessels in your area.

Retrieving a man overboard is the most difficult
step in the rescue. (K. Byers photo)

Search patterns depend on wind and sea conditions. If you have a plotter, use it.

You may decide to approach the person from **his** leeward side. This does not create a lee for the person, but it may make communication easier because most people naturally float with their backs to the wind and waves. Or, you may prefer to keep him on **your** leeward side, but this can be dangerous in rough seas. Regardless of your approach, keep the propeller away from him. Ultimately, the approach depends on the sea conditions, your retrieval method, your vessel's maneuverability, and the victim's condition.

4) Retrieve the person. This is often the most difficult step of all. Its ease depends both on how much the person in the water is able to help and whether you have something to help get him back on board. This is where earlier practice pays off.

Putting a crew member in the water to help retrieve the person should be done only as a last resort. If attempted, the in-water rescuer must be in a survival suit and attached to the vessel by a lifeline. Otherwise you might have two crewmen to rescue.

5) Gently treat the person for drowning, hypothermia, and other injuries as needed. See the First Aid chapter for details.

6) If the person is not immediately located, notify the Coast Guard and other vessels in the vicinity, and continue searching until released by the Coast Guard.

Do all you can to keep yourself on board. It's much easier to take precautions than to get rescued when you fall overboard.

VII

Sea Survival

"Fishermen OK After Boat Sinks"

"KODIAK (AP)—Two people who abandoned a sinking fishing boat early Wednesday were rescued unharmed near Kodiak, the Coast Guard said.

"The communications center at the Kodiak Coast Guard base heard a distress call from the 38-foot *Mary L* and dispatched a helicopter to the scene about 5 miles from Broad Point. [The two men], both of Kodiak, had donned survival suits and gotten into a life raft.

"The helicopter spotted a flare fired from the raft and hoisted the men aboard." (*Daily Sitka Sentinel*, Sitka, Alaska, October 25, 1989.)

"Fishing Boat Sinks; Crew OK"

"The 56-foot fishing vessel *Unimak* capsized and sank near Icy Bay early Saturday morning, but the four persons aboard were rescued from a life raft shortly afterward, the Coast Guard reported.

"The Coast Guard said the *Unimak* was departing Icy Bay, about 50 miles west of Yakutat, around 6 a.m. when it was rolled over by a large wave. It happened too fast for a mayday call to be sent out, but the sinking activated the boat's EPIRB, an emergency radio signaling device, setting a Coast Guard search into motion." (*Daily Sitka Sentinel*, April 18, 1988.)

"In another incident, the 50-foot seiner *Rebecca*, from Juneau, sank in Ernest Sound on the east side of Clarence Strait Sunday morning, but all four people aboard were rescued.

"The four were in the water a short time before being picked up by the F/V *Marysville*, and then transported to Ketchikan aboard a 41-foot Coast Guard boat from Ketchikan.

"The men had three survival suits and a small raft, and managed to get off a mayday call before their boat went down." (*Daily Sitka Sentinel*, September 12, 1988.)

Fishing is hazardous work. Incidents such as those related above happen year-round in the North Pacific. Hardly a week goes by without a news story describing the sinking of a fishing vessel. When you carefully read the accounts, a common thread emerges: (a) the survivors usually had immersion (survival) suits and/or life rafts, (b) they sent a Mayday or had an EPIRB (Emergency Position Indicating Radio Beacon), (c) they knew enough about their survival gear to make it work for them, and (d) they had the will to live.

Do you know how to survive the loss of your fishing vessel? What follows are proven procedures that will increase your ability to survive the loss of your ship at sea.

Abandoning Ship

Some emergencies occur quickly with little or no warning, while others may have a delayed onset. If you are not sure you can control a situation, contact the Coast Guard and ask them to set up a call schedule and standby. They will know something is wrong if you don't call in on schedule and will then begin a rescue.

Flooding, fire, capsizing, or grounding may someday make your vessel unsafe. Abandoning ship may become necessary, but it is a serious move, especially in North Pacific waters.

A tragic example of an inappropriate order to abandon ship occurred aboard the F/V *St. Patrick* several years ago. During heavy weather the captain ordered all hands off the vessel. Ten crew members died trying to survive in the rough seas. When rescuers arrived, they found the *St. Patrick* still floating—with immersion suits left on board. The vessel, a grim reminder of a premature order to abandon ship, was towed to Kodiak.

Make sure you abandon ship **only** when you are certain that being on board the vessel is more dangerous than being in the water. Once the decision has been made to leave the vessel, there are three things to do: broadcast a Mayday, prepare the crew, and inflate your life raft.

MAYDAY

Make sure your radio is on and you transmit on channel 16 VHF or 2182 kHz SSB or 4125 kHz SSB. Then state:

1) MAYDAY, MAYDAY, MAYDAY.

2) Your vessel's name/call sign **three** times.

3) Position (latitude/longitude and loran are preferred).

4) Nature of distress (fire, grounding, medical emergency, etc.) .

5) Number of people on board (P.O.B.).

6) Amount and type of survival gear on board (immersion suits, life rafts, EPIRBs, flares, etc.).

7) Vessel description (length, color, type, etc.).

8) Listen for a response. If there is none, repeat the message until it is acknowledged or you are forced to abandon ship.

If you are fortunate, a nearby vessel will hear your Mayday and pick you up.

Crew Preparation

Before abandoning ship, put on your shoes or boots, a hat, and several layers of wool, polypropylene, or other clothes that will help keep you warm even when they are wet. Then put on your immersion suit. If you can't wear shoes or boots in your suit, stuff them inside the suit. You'll need them on shore.

Make sure your personal survival kit is inside your suit or the suit's pocket. (See Chapter I for information on this kit.)

Cautions

It can be difficult to escape out of a flooded compartment if you are already wearing flotation, so take care not to get trapped inside the vessel.

Don't get snagged by the rigging. A vessel that is being abandoned is often a tangle of spars, lines, fishing gear, and other paraphernalia. Give yourself a clear pathway to escape, and beware of getting tangled or trapped.

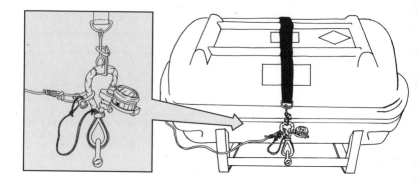

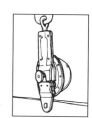

Release the pelican hook to free the life raft. Older rafts may have the type of hydrostatic release shown at right. Hit the button to free the raft.

Inflate the Life Raft

1) Free the canister from its cradle. Some models have a pelican hook that you undo; others are freed by hitting the brass hydrostatic release.

2) Carry (do **not** roll) the canister to the vessel's **lee** side.

3) Check to make sure there is nothing below you in the water that could damage the raft, then throw the canister overboard. (One raft manufacturer recommends inflating his raft on deck. Read the information that comes with your raft.)

4) Pull on the painter until the raft inflates. You will need to pull up to 200 feet of line and then—when you feel resistance—give a sharp tug to inflate the raft.

If the vessel sinks before you can remove the raft from its cradle or inflate it, the hydrostatic release will activate at a depth of approximately 15 feet, freeing the raft from its cradle. (A few life raft models are designed to float free out of their cradles without any hydrostatic releases.) As the raft floats toward the surface, and the vessel continues to sink, the painter triggers the raft's CO_2 cartridge, and the raft inflates. Continued pressure on the raft's painter from the sinking vessel will cause the painter's weak link to part, separating the raft from the vessel.

If the raft is not inflated when it hits the surface, pull on the painter to trigger the CO_2 cartridge, but watch out for the bands that hold the raft together. They can pop off with considerable force.

Leaving the Vessel

Make sure your EPIRB is either in the raft or with you before you abandon ship.

Abandon ship safely and **stay as dry as you can**. If you are not able to step directly into the raft, ease yourself into the water and use the painter to get to the raft. Try to avoid jumping into or on the raft. This can be dangerous, especially if it is done from a height or with people inside. Injuries will only make your survival situation more difficult.

If you must jump into the water, store your eyeglasses unless you know you can keep them on, and use the following four steps to help protect your head, neck, and groin:

If you must jump into the water, take precautions to protect your head, neck, and groin.

1) Stand on the vessel's edge (the lee side is usually best) so you can step sideways off the vessel.

2) If you are wearing an immersion suit:

   - Make sure it is fully zipped and the face flap is secured.

   - **Do not** blow up the suit's air bladder until you are in the water. (This prevents possible neck injuries.)

   - Put your arm that is nearest the vessel over your head, and insert the thumb on your other hand into the suit's hood. This will protect your head and permit air to escape from the suit when you hit the water.

   If you are not wearing an immersion suit:

   - Protect your head with your hand nearest the vessel, and use your other hand to cover your nose and mouth.

3) Look down to make sure there is no debris in the water.

4) Step off the vessel and cross your legs at the ankles. Once you hit the water you will momentarily submerge, but will then bob up.

Although this jumping method works, the best way to enter the water is still to ease yourself in.

In the Water

Without Survival Suit or Life Raft

This is not a good situation! As you read this, think about where your suit and life raft are stored. Can you get to them if the vessel capsizes? Don't count on being able to go below decks for your suit. Take the time **now** to make sure your raft will float free and your suit is accessible at your work place or near a vessel exit.

If, however, you do find yourself in the water with no suit or raft:

1) **Stay with the boat** as long as you can, even if it is overturned. This will keep you drier and make you more visible to rescuers.

2) **Stay warm and dry** by keeping your clothes and boots on (they will **not** make you sink) and getting as much of yourself out of the water as possible. Your body cools 25 times faster in water than in air of the same temperature. If you cannot get out of the water, assume the Heat Escape Lessening Posture (H.E.L.P.) or huddle with other people.

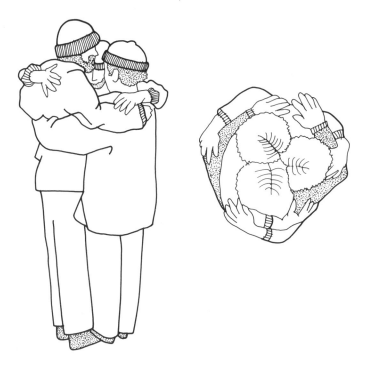

Use the huddle position in the water.
Side view (left) and top view (right).

3) **Stay afloat** by wearing a PFD and hanging onto floating debris, logs, or other objects.

4) **Stay as still as possible**. Cold water moving against your high heat loss areas—head, neck, armpits, sides of the chest, and groin—steals your body's heat. Staying still can extend your in-water survival time by up to 30 percent.

5) **Signal for help** as soon as you can. (Another good reason to carry a personal survival kit.)

With an Immersion Suit without Life Raft

If you have abandoned ship with your immersion suit but did not have a chance to put it on, get into it as soon as possible. (See the PFD chapter for information on donning the suit in the water.)

Improve your shelter by blowing up the suit's air bladder as soon as you can. This will help raise your body higher out of the water—an especially important factor in rough seas. Increase your buoyancy and visibility by hanging onto the capsized vessel, buoys, or other floating objects.

Signal for help with your EPIRB by turning it on and leaving it on **all** the time. If you have a PFD light, and it is not automatically activated, turn it on at night. If you have more than one flare or orange smoke, fire one as soon as you can, but leave the others for when rescuers are in sight. When firing flares, hold them well away from you; the hot drippings can burn through your suit. Keep the wind at your back, aim 60 to 85 degrees up from the horizon (aim higher in windy conditions), and turn your head away just before you fire. Assist other survivors and stay together to boost morale and be a larger signal.

If you need to swim, do it on your back to help keep water out of your suit. If the seas are calm, you can hook arms and legs with another survivor and propel yourself fairly quickly.

Some survivors have complained of being too hot in their immersion suits. Although this is unlikely to be a long-term problem in North Pacific waters, the excitement of abandoning ship can make you feel hot, especially if you have several layers of clothing on under your suit. Even if you are tempted to open the suit to cool off, **don't.** It is important to stay as dry as possible. Getting soaked from sweat isn't good, but it is better than having water in your suit.

Keep your spirits up! Several people have been rescued after floating for over 20 hours in an immersion suit.

Stay together by tandem swimming.

The Life Raft

Righting the Raft

Upside-down rafts need to be righted. First, find the righting strap—a nylon strap on the underside of the raft—and make sure you are on the **same** side as the CO_2 bottle (so the bottle won't come crashing down on your head). The newer model rafts have a notice stenciled on the bottom of the raft or pontoon that says, "RIGHT RAFT THIS SIDE" to let you know which side to work from.

Then, pull yourself up onto the bottom of the raft with the righting strap and lean back. Be careful not to tangle your hand or arm in the righting strap. As the raft flips over, hold one hand over your head to push it off you.

If you have trouble righting the raft, use the wind and waves to your advantage.

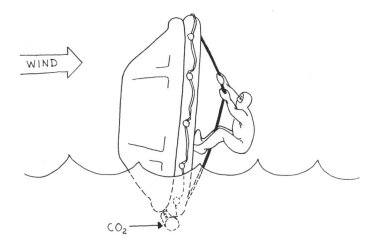

Right the life raft by hanging on to the righting strap and leaning back.

Boarding the Raft

Getting into the raft from the water takes effort, especially if you are tired, cold, or injured. Some newer model rafts (SOLAS approved) have an inflated boarding platform that makes entering the raft easier. (SOLAS, Saving of Life at Sea, is an international group formed after the sinking of the Titanic. It sets standards for marine safety equipment.) If there is no platform, locate the web boarding ladder and use your suit's buoyancy and the raft's straps and ladder to your advantage. To board, grab as high onto the ladder or the raft as you can, and crouch low in the water. Then, all at once, pull yourself up with your arms, and kick with your legs. Your suit's buoyancy should help push you up out of the water. Boarding a life raft is an excellent example of a skill that gets easier with practice.

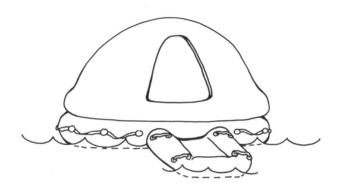

Life raft with boarding ramp.

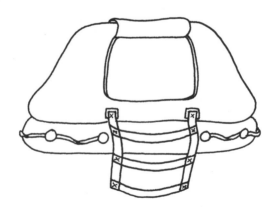

Life raft with ladder.

Life Aboard the Raft

Once you are aboard the raft, it is a whole new survival situation. It will help to apply the Seven Steps to Survival:

1) Recognition
2) Inventory
3) Shelter
4) Signals
5) Water
6) Food
7) Play

Although the Seven Steps are listed in order of priority, you may be able to do more than one step at a time.

Recognition

Recognize that you are not out of danger, but that your survival situation has changed.

- Stay tied to your vessel as long as it is safe. When you need to cut yourself free, use the raft's knife (located near the door) to cut the raft's painter. (Be careful with knives around the raft.)

- In very rough seas or high winds, too many people in one part of the raft could cause it to flip, so keep the crew's weight evenly distributed over the raft.

Inventory

As you inventory your situation, equipment, and the crew's condition, think about what can help and hurt you.

- Provide first aid for seriously injured or hypothermic persons. Less serious injuries, such as sprains or minor cuts and burns, can be treated once you have gone through the Seven Steps.

- Open the life raft's equipment pack and **tie everything to the raft**—including the paddles. Not doing this is a common mistake among survivors, and is **always** regretted.

- Take seasickness medications as soon as possible and **before** people get seasick. Even the most seasoned fishermen have been known to get seasick in a life raft.

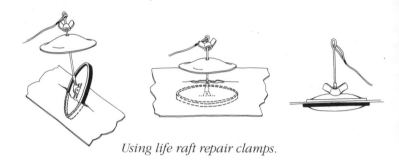

Using life raft repair clamps.

- Inventory everything you have, including what people have in their pockets and immersion suits. This may be a verbal inventory until the canopy is up, the doors are closed, the raft is bailed and dried, and the floor is inflated. Make sure sharp objects cannot puncture the raft.
- Inspect the raft for damage, and repair it as needed using the kit in the raft's equipment pack.
- Stream the sea anchor. This will reduce your drift rate, helping to keep you near your last reported position, and will also help prevent capsizing in heavy seas.
- Gather up useful floating objects. Be creative, and don't throw anything away unless it will hurt you or damage the raft.

Streaming the sea anchor will slow your life raft.

Equipment Required on USCG-Approved Life Rafts

Equipment	Coastal Pack	Solas B Pack	Solas A Pack
Quoit and Heaving Line	1	1	1
Knife (Buoyant Safety)	1	1	1
Bailer	1	1	1
Sponge	1	2	2
Sea Anchor	1	2	2
Paddles	2	2	2
Whistle	1	1	1
Flashlight with spare batteries and bulb	1	1	1
Signal Mirror	1	1	1
Survival Instructions	1	1	1
Immediate Action Instructions	1	1	1
Repair Outfit (1 set sealing clamps or plugs)	1	1	1
Pump or Bellows	1	1	1
Tin Openers	0	0	3
First Aid Kit in Waterproof Case	0	1	1
Rocket Parachute Flares	0	2	4
Hand-Held Flares	0	3	6
Buoyant Smoke Signals	0	1	2
Copy of Life-Saving Signals	0	1	1
Fishing Tackle	0	0	1
Food Ration	0	0	2,378 calories/person
Water	0	0	1.5 liters/person
Rustproof, Graduated Drinking Vessel	0	0	1
Anti-Seasickness Pills	0	6/person	6/person
Seasickness Bag	0	1/person	1/person
Thermal Protective Aid	0	Enough for 10% of persons or 2, whichever is greater	Enough for 10% of persons or 2, whichever is greater

Note: The quantity of each item may change as regulations change. Know what is packed in **your** raft. Within strict space limits, other small items such as a Class B or S EPIRB, medication, eyeglasses, etc., can be specially packed in your raft.

Shelter

Do as much as possible to conserve your body's heat.

- Put up the canopy.
- Inflate the floor with the air pump.
- Bail out the water.
- Close the raft's doors to raise the inside temperature of the raft.
- Top off the buoyancy tubes as needed. The CO_2 in the tubes loses volume as it cools.
- Empty water out of immersion suits and wring out wet clothes. (You may not be able to do this if it is too rough.)
- Inspect the raft for holes or areas of wear at least once a day, and pump up the buoyancy tubes as needed.

Signals

Make sure your signals both attract attention and convey your need for help.

- Turn on the **EPIRB** and leave it on until you are rescued. The EPIRB should be securely attached to the raft and placed outside because some canopied rafts have radar reflecting material in them that may prevent radio signals from getting through the canopy.

EPIRBs

EPIRBs are similar to Emergency Locator Transmitters (ELTs) carried in aircraft. They emit a signal that can be picked up by an aircraft or satellite, and then transmitted to a ground station to begin the search and rescue. The nearest ground stations in the North Pacific are in Kodiak, Alaska, and Vladivostok, Russia.

The COSPAS/SARSAT (Cosmicheskaya Sistyema Poiska Avariynych Sudov/Search and Rescue Satellite-Aided Tracking) satellite system is an international search and rescue effort. Russia and the United States own the satellites, Canada developed the repeaters, and the processors are French. Bulgaria, Denmark, Finland, Norway, Sweden, and the United Kingdom also support the system.

Emergency Position Indicator Radio Beacons (EPIRBS)

Type	Frequency (mHz)	Locating Accuracy (miles)	Minimum Signal Life	Self-Activated?[3]	Float Free?	Comments
Category 1	406 121.5[1]	1-3	48 hrs. at -20 C	Yes	Yes	Each unit digitally coded to its owner—be sure to mail in your registration card. Satellite can receive and hold 406 mHz signal, so signal is picked up quickly. Global coverage. Expensive and sophisticated.
Category 2	406 121.5	1-3	48 hrs. at -20 C	No	No	See Category 1 comments.
Class A	243[2] 121.5	5-20	48 hrs. at -20 C	Yes	Yes	Both frequencies sent to satellite, but ground station must be in sight of satellite to relay signal. This can delay signal pickup. Coverage is not global.
Class B	243 121.5	5-20	48 hrs. at -20 C	No	No	See Class A comments. Available in small size to fit in PFD pocket.
Class C	VHF 16 (alerting) 15 (homing)	20, or line of sight	24 hrs., then shuts off. Can be reactivated.	No	No	For coastal use only. Signal may be picked up by other vessels, USCG, or search & rescue units if channel 16 is not busy. Not desirable in most Alaska waters.
Class S	243 121.5	5-20	48 hrs. at -20 C	Yes, with raft inflation	No	See Class A comments. Optional for life rafts. Can only be checked when raft is inspected.

[1]121.5 mHz is civilian aviation distress frequency. [2]243 mHz is military distress frequency. [3]Batteries must be good, and properly connected.

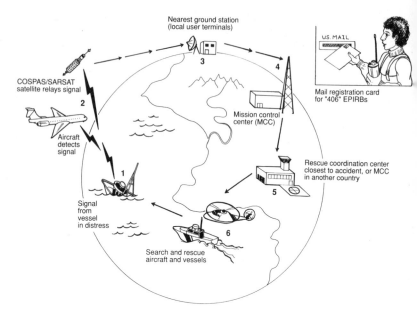

How an EPIRB signal is picked up and relayed.

The newer Category 1, 406 mHz EPIRBs are more effective than other types because they have a higher power output and improved frequency stability. In addition, the satellites can hold the 406 mHz signal information until a ground receiving station is in sight, something they cannot do with the 121.5 or 243 mHz signals. Both the signal and ground receiving station must be in sight to relay the 121.5 and 243 mHz signals. This added power, improved frequency stability, and signal holding capability gives the 406 mHz EPIRB much broader—in fact, global—coverage, which other EPIRBs do not have.

- Fire one **signal flare** as soon as possible, but keep others for when rescuers are in sight. Treat flares like a gun, and never point them at anyone. Hand-held flares must be held out over the water on the downwind side at an angle to prevent hot "drippings" from burning either you or the raft. Burning rafts make bright signals—but they don't float!

*When firing a flare, hold the flare away
from the raft, and turn your head away.*

- Carefully study the directions for each signal and make sure each person in the raft knows **how** and **when** to use each type of signal.

- Keep other signals available for when rescuers are in sight. It may help to divide the signals into day and night types, keeping the appropriate ones near the lookout, ready to be deployed at a moment's notice. Some flares work from both ends, but don't trigger them both at the same time, of course. After firing, be sure to douse the end you have used and to save the other end for later.

- Post a lookout to spot ships, aircraft, land, and useful debris; to signal rescuers; and to listen for aircraft and surf. A routine of lookout watches is important to help establish a sense of being more in control of the situation and to maintain some sort of command structure.

Orange Smoke, Flares, Signal Mirror, and Dye Marker

Type	Optimum Visibility*	Signal Duration	Advantages	Disadvantages
Orange Smoke	3-5 miles at water level, more from the air	50 sec.- 2 min.	Compact, good for day use, can show helicopter pilots wind direction, can help locate a person overboard in daylight.	Smoke dissipates rapidly in windy conditions, must be used in a well-ventilated area, container can damage raft or cause personal injury, outdated containers have high failure rate.
Hand-Held Flare	3-5 miles	50 sec.- 2 min.	Compact, longest burning of any flare type, secondary use as a fire starter, inexpensive. Helps rescuers locate you.	Ash and slag can cause injury and damage raft, signal is low to the water, high failure rate with outdated flares.
Meteor (Aerial) Flare	10-20 miles	5.5-8 seconds	Compact, helps alert rescuers.	Ash and slag can cause injury and damage raft, high failure rate with outdated flares, can be difficult to operate with cold hands.
Pistol-Fired Meteor Flare	19-40 miles	5.5-30 seconds	Easy to use, helps alert rescuers.	High failure rate with outdated flares, flares unusable if pistol breaks, flare can cause personal injury or damage raft.
Rocket-Propelled Parachute Flare	40 miles	30-60 seconds	Most visible flare on the market for night use.	High failure rate with outdated flares, flare may drift or be blown away from your area, can cause personal injury or damage raft.
Signal Mirror	40 miles	As long as there is enough light.	Compact, easy to use, good for day use, doesn't deteriorate in bad weather.	Needs sun or other light source to work, must be manned continually.
Dye Marker	10 miles at 3,000 feet	Dye weakens in 20-30 min. in calm seas, dissipates sooner in rough seas.	Easy to use and carry, can also be used on snow. Does not deteriorate in bad weather.	Only visible during the day, not as visible by sea as from the air, dissipates rapidly in rough seas.

*Actual visibility depends on weather, altitude of rescuer, and whether it is day or night.

- To operate a **signal mirror,** position yourself with the mirror, sun, and potential rescuers in front of you. Using one hand as a sight, locate the rescuers and aim the mirror's reflected light through the sight. Polished metal will also work as an improvised signal mirror.

  Sweep the horizon often with your mirror's signal—you may attract the attention of someone you cannot see with your naked eye. On a sunny day, the mirror's reflected light can be very bright, so take care not to blind potential rescuers, especially pilots.

- Arrange for other duties and watches.

- Check to make sure the raft's outside canopy light works.

- If you have a VHF radio, transmit a Mayday on channel 16. Consider your location to help determine how often you transmit.

- Conserve the flashlight batteries by only using the light when it is absolutely needed.

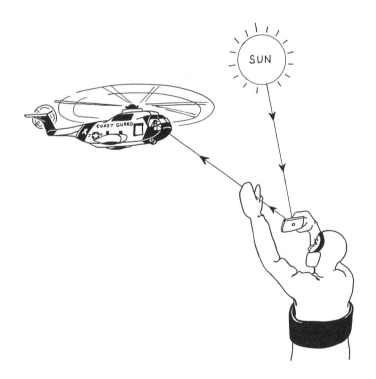

Aiming a signal mirror.

Water

You need a safe source of drinking water.

- Whenever possible, gather and store rainwater by using the raft's water collection system.
- Decide on a daily water ration for each person.
- **Never** drink sea water or urine.
- Vomiting dehydrates you, so continue to take seasickness medications as directed on the package.

Food

Most people can live for weeks without food as long as they have water to drink.

- Do not eat if water is not available. Eating without drinking water accelerates dehydration.
- Food helps lift spirits, so if it is available, decide on food rations. Captain Bligh wrote about a method of dividing food that was time-honored even in 1789: "I divided it [a noddy, about the size of a pigeon], with its entrails, into 18 portions, and by a well-known method of the sea, of 'who shall have this' it was distributed, with the allowance of bread and water for dinner. . . . One person turns his back on the object that is to be divided; another then points separately to the portions, at each of them asking aloud, 'Who shall have this?' to which the first answers by naming somebody. This impartial method of division gives every man an equal chance of the best share."

Play

Keep busy and take measures to improve morale.

- Focus on constructive ways to improve your situation.
- Practice "dry runs" with your signals, catch drinking water, clean up the raft, stay positive!
- Tell jokes. This is where practice can really make a difference.
- Try your luck at fishing.
- Your brain is your best survival tool. Be creative!
- Think like a survivor, not like a victim.
- Do not give up.

Prolonged Sea Survival

When Nature Calls

Avoid the urge to urinate and defecate in your suit; urine and feces are irritating to the skin and can cause body sores. Instead, use plastic bags and empty the contents outside the raft.

Do not be surprised if your bowel action stops because of short rations and the lack of activity. This is not unusual.

Water

Maintaining your body's water balance is one of the prime requirements for survival. You can last weeks without food, but not much more than a week without water. The good news is that frequent rain in most areas of the North Pacific will permit you to gather water with the raft's water catchment system.

If you do not have water, don't eat. Digestion uses water.

Thirst can be a problem and it helps to remember that it is not always due to water need. The sensation of thirst can be created by sugar and salt—and even by sweetened beverages. So when water is scant, avoid such food and drink. Thirst may be reduced by chewing gum—or practically anything as long as it's not saltwater soaked—but this relief does not reduce the body's need for water.

Don't drink sea water, urine, or alcohol even if your water supply is limited. They all lead to dehydration. Sea water has a salt content of 3.5 percent, the equivalent to a full teaspoon of salt in a six-ounce glass of water. Drinking sea water exaggerates thirst and promotes water loss through the kidneys and intestines, shortening your survival time. The toxic waste products in urine add to the agony of thirst, contribute to dehydration, and lead to a body temperature of 105°F and above. Alcohol, too, promotes water loss through the skin and kidneys. (The thirst and dry feeling experienced during a hangover is from dehydration.)

Every bit of body water you conserve will increase the length of your survival. Much of your body's moisture is lost through breathing and sweating, so try to avoid unnecessary exertion. Ration your sweat, and save your energy until you truly need it. Don't try to paddle your raft upwind or upcurrent.

If you are completely without water, you are apt to get delirious in about four days. If someone becomes delirious, it may take physical force to keep him aboard.

Flipped Over

Supplies can be lost out of a raft in both calm and rough seas. Secure all items!

If the seas get rough and the wind howls, use human ballast to help keep the raft from flipping. Keep your weight low in the raft, anticipate the waves and, at the proper time, shift your weight toward the waves.

If you find yourself ejected out of or tumbling inside a capsized raft, your means of righting it will depend on the sea state and number of crew members that can help. The people inside may be able to right it by crawling up onto the floor and using the wind to help flip it over. An outside person may be able to assist by using the righting strap (he should stay on the same side as the CO_2 bottle). When a raft flips, don't get separated from it. A surge of energy and determination not to give up will help immensely.

Psychology of Survival

Are some people more likely to survive an emergency than others? Survival is determined to a large degree by how people react to their emergency. Before discussing the actions and attitudes that lead to survival, let's cover some of the general reactions.

A percentage of survivors—estimated to be less than 10 percent—feel calm during an emergency. Being calm can help you make the right moves, but being too calm can be dangerous, especially if it leads you to inaction or a failure to acknowledge the emergency. Fear, likewise, can be healthy if it motivates you to action, but too much fear can be deadly. Panic occurs less frequently than is commonly believed, but when it does occur it is very contagious. It has been proven that preparation and training help decrease fear and panic.

Initial reactions to an emergency can also include denial, and feeling numb, stunned, or bewildered. Some people may be in psychological or physical shock, while others may exhibit inappropriate behavior such as searching for a flashlight instead of taking action to rescue someone. Still others may become hyperactive, doing much but accomplishing little.

During a survival situation, emotions may change from day to day or even hour to hour.

Anger—both at companions and rescuers—is a common feeling. The participants in one life raft **drill** were ready to fight after only eleven hours in a raft. In a real emergency, tempers often flare and accusations fly after unsuccessful attempts to signal rescuers. Anger is not surprising, considering the cold, cramped, wet conditions of a life raft, but sustained anger can be deadly. A good rule is to resolve your anger before the day ends.

While some survivors experience rage, others may be totally passive and unable to help themselves. Although passivity can be a psychological reaction to the emergency, it can also be brought on by seasickness, hypothermia, injury, lack of prescribed medications, hunger, and thirst. If the cause is psychological, some people will perk up when asked to perform simple useful tasks such as bailing or keeping watch, or when directed to help an injured companion.

Some people experience guilt about what they did or did not do, especially if they think they contributed to the disaster. This can be a debilitating emotion if allowed to continue. Guilt does not help resolve the survival dilemma.

The suicidal impulse is no stranger among survivors. The crew member who suddenly starts over the side, saying, "I'm going down to the corner for a glass of beer," is suffering from hallucination and disassociation of time and place. It is your duty to restrain him.

Survivors tell us a great deal about why they survived. Whether their ordeal involved drifting in a life raft, being shipwrecked on land, or being held as a prisoner of war, there are common themes that run through their stories. Some of them read like headlines:

• ACCEPT YOUR SITUATION BUT DON'T GIVE IN TO IT.

• ACT LIKE A SURVIVOR, NOT LIKE A VICTIM.

• DON'T GIVE UP.

• BE POSITIVE.

• PRAY.

Survivors report that it is important to try to regain some sense of control over your situation, especially by acting to improve your circumstances. Schedules and routines can also help.

When you don't think you can climb back into the raft after it has flipped for the sixth time, or you don't think you can stand another day, **don't give up**. Live your ordeal one hour or one minute at a time if necessary. Remember your family and friends, and concentrate on returning home to them. Plan your future.

Be positive by talking about when, not if, you will be rescued. Of course, if you have filed a float plan, sent a Mayday, or turned on your EPIRB, it is easier to be positive. Remember the seventh step: Play.

Many survivors describe the power of prayer in an emergency. Don't underestimate it.

Do not downplay the role your emotions can play. You can do many things to help yourself survive an emergency. Force yourself to stay on your team. Find the will to live!

Rescue!

After you have abandoned ship, there is hardly a more exciting sight than a vessel or aircraft headed your way. Since a raft is difficult to see from sea or air at any substantial distance, keep signaling until you are sure you have been sighted, but then stop. Don't keep firing flares at rescuers after they have located you. Most pilots and skippers take a dim view of flares fired at close range.

An aircraft will clearly show when he sees you—perhaps by "buzzing" you or dipping his wings. However, an aircraft that has spotted you may have to leave for periods of time due to weather, lack of fuel, or darkness. Sit tight and save further signals for his return. Many people become depressed after a rescue craft has sighted them, but then had to leave for some reason. Don't fall into this trap.

If you are approached by a Coast Guard helicopter, make sure all your gear stays securely lashed. Pilots don't like objects flying up into the aircraft's rotors, nor are they especially happy when someone secures the helicopter's trailing line. Would you want a raft tied to your flying machine?

The U.S. Coast Guard may use a rescue swimmer to assist you. These aviation survival men and women drop into the water and swim to the raft or person in the water. The chopper's nearly 100-knot rotor wash may make communication difficult, so pay close attention. You will be hoisted into the helicopter in a basket or litter. Keep your body—especially your hands—inside the lifting device. One person will be hoisted at a time.

Beaching

Your chance of survival is much greater on land than on water, so if you haven't seen any potential rescuers and you are near shore, try to get there. You will need to judge the beaching situation to determine your best course of action. If you go ashore in just your immersion suit, try to head in feet first.

As you drift toward shore, you may encounter high surf, rocks, cliffs, strong currents, and floating logs. Life rafts have limited maneuverability, but try to come ashore in the least hazardous area. Use the paddles in the raft's survival kit to help you through the surf.

Your raft makes a great shelter and signal, so keep it if you can. Shore survival is discussed in the next chapter.

Reactions to a Traumatic Situation

After your rescue, you may find that your ordeal is still not over. Some people who survive a traumatic situation continue to feel the stress of the event afterward. This is as true for vessel sinkings as it is for assaults, hurricanes, tornadoes, war, shootings, and other emotionally devastating events.

Common reactions may include questioning why you lived and other people died, or wondering what you could have done differently. People who experience post-traumatic stress often have flashbacks, nightmares, and difficulty sleeping. Depression, irritability, fear, mood swings, and loss of appetite are also common. All of these reactions are normal.

Once you have physically endured the experience, how can you survive it emotionally? Three things can help considerably: accept your feelings, talk to others, and try to find something positive in the experience. All three are much easier to say than do, especially if you have a difficult time dealing with emotions or talking. People who have the hardest time recovering are those who refuse to talk about what happened. It is best to talk with people who can understand how you feel and who will really listen to you. Your friends, pastor, or local mental health professional may be able to help you by listening.

Trying to find something positive in what happened may sound trite, but it helps. Some survivors buy more survival equipment and take training. Others, through their personal testimony or quiet talk, spread the word that being prepared and having the will to live increases your chances of survival.

Shore Survival

Most North Pacific fishermen who abandon ship in immersion suits and life rafts never face a shore survival situation. Their ordeal occurs at sea, and most are rescued within two or three days.

Those who end up on shore must find ways to stay warm, signal for help, and gather water and food. This takes resourcefulness and a fierce determination to survive, but it can be done. Two examples follow:

- A crewman from a capsized 32-foot gillnetter survived on Nunivak Island for nearly two weeks before discovery and rescue. He ate kelp and year-old berries to sustain himself.
- When the F/V *Master Carl* capsized in Prince William Sound, all four men abandoned ship into a life raft. During the night the raft flipped in huge seas, but they managed to right it and climb back aboard. The two men who ultimately survived were separated from the others when the raft was pounded by 20-foot surf while approaching an island. When they were ashore and sleeping under the raft, an inquisitive grizzly bear convinced them that coming ashore hadn't eliminated danger—only changed it. After the bear retreated, they battled exhaustion and hypothermia before being picked up by a Coast Guard helicopter.

Seven Steps to Survival

Unless you find yourself washed ashore near a house or cabin, it will take serious action to survive being stranded on the shores of the North Pacific. Prioritize your actions with the Seven Steps to Survival:

1) Recognition
2) Inventory
3) Shelter
4) Signals
5) Water
6) Food
7) Play

Recognition

Recognize that your situation has changed and that you face challenges different from those at sea. Your survival depends to a large degree on your state of mind; think and act like a survivor.

Inventory

Take time to inventory your circumstances and resources, and to determine what is working for and against you. Where are you? How is the weather? What skills do you and your fellow survivors have? Is there useful debris on the beach? How is the crew's physical and emotional state? Is anyone injured or hypothermic?

Be creative as you think of items for shelter, signals, water, food, and play. Don't throw things away; you may need them later. A torn piece of plastic can become a rain jacket, part of a shelter, a water collector, or bag for food. A belt can be an essential part of a shelter, its buckle a reflective signal.

Shelter

Your biggest enemy in a shore survival situation along the North Pacific is not starvation, dehydration, or bears. It's the quiet killer— hypothermia, the lowering of the body's core temperature.

The best way to prevent hypothermia is adequate protection from the elements. In an emergency, you need to be concerned with finding or making shelter, including a hat. Help yourself stay warm by keeping your head covered; you lose 50 percent of your body's heat through your head.

Your emergency shelter may be your clothing, your immersion suit, your life raft, piles of moss and leaves, or a shelter that you construct. To be effective, shelter must be weatherproof, insulated, and small.

- **Weatherproofing** that protects you from the wind, rain, and snow is a must. Wind can rapidly make you feel colder than the actual air temperature by taking away the warmer air near your body. The wind chill chart clearly demonstrates this point. Water—whether it's fresh, salt, or from sweat—robs your body of heat 25 times faster than air of the same temperature. Get dry and stay dry.

WIND CHILL CHART
WHAT THE THERMOMETER ACTUALLY READS

WIND SPEED (MPH)	50	40	30	20	10	0	-10	-20	-30	-40	-50	-60
CALM	50	40	30	20	10	0	-10	-20	-30	-40	-50	-60
5	48	37	28	16	6	-5	-15	-26	-36	-47	-57	-68
10	40	28	16	4	-9	-21	-33	-46	-58	-70	-83	-95
15	36	22	9	-5	-18	-36	-45	-58	-72	-85	-99	-102
20	32	18	4	-10	-25	-39	-53	-67	-82	-96	-110	-124
25	30	16	0	-15	-29	-44	-59	-74	-83	-104	-113	-133
30	28	13	-2	-18	-33	-48	-63	-79	-94	-109	-125	-140
35	27	11	-4	-20	-35	-49	-64	-82	-98	-113	-129	-145
40	26	10	-6	-21	-37	-53	-69	-85	-102	-116	-132	-148

WHAT IT EQUALS IN ITS EFFECT ON EXPOSED FLESH

DANGER

GREAT DANGER

|← LITTLE DANGER IF PROPERLY CLOTHED →|← DANGER OF FREEZING EXPOSED FLESH →|

- **Insulation** helps keep your body heat from escaping. The warmest shelters use trapped, still air for insulation.

- Build a **small** shelter that will trap your body heat and slow further heat loss. A large shelter may keep wind, rain, and snow away, but it will do a poor job retaining your body heat. Build your shelter so its ceiling and sides are no more than six or eight inches away when you are lying down inside. If you have a tendency to get claustrophobic, you may need to make your shelter a little larger or sleep with your head near the door.

Shelter Construction

Your emergency shelter's location, size, and shape will be affected by your circumstances, resources, imagination, the number of survivors, and your energy levels.

If there is little time to build before dark, your first night's shelter may be a debris bed made from a pile of moss, grass, leaves, and branches. Make the pile as high and as weatherproof as you can, and crawl into it for protection.

Building a more permanent and warmer shelter will take several hours, so you may want to build some signals first. Do not totally ignore your shelter needs just to build signals, however.

Look for the natural beginnings of a shelter. You may be able to build in a cave, or next to a fallen log or overhanging cliff. When you choose a site, make sure ground water or rain won't collect in your shelter. Select a location that is protected from the wind, rain, and snow, yet close to construction materials, signals, and a source of water. Balance your need to be protected from the elements with your need to be visible to rescuers.

Begin your shelter with a bed, using piled-up branches, small driftwood, and other materials to raise you off the cold ground. Then lay down a two to three foot deep insulating layer of grass, leaves, moss, and small branches. (It will compress appreciably with use.) A thick, dry, insulating layer underneath you is critical, so use plastic or broad leaves as a waterproof layer on top of the insulation.

Once you have completed your bed, begin the structure. Use driftwood, deadfall logs, or small branches for support beams. Place them close enough to support the insulation and waterproofing layers on the shelter's ceiling and sides, and arrange all branches so the water runs off your shelter. Keep your shelter small. A small, well-insulated shelter can be warmed by your body heat much easier than a big one.

Beginning a shelter.

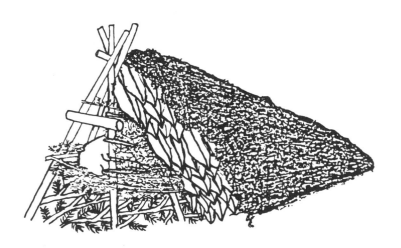

*Use plastic, bark, leaves, or branches as
roofing material for your shelter.*

How you insulate and waterproof your shelter depends on your materials and ingenuity. Plastic will help keep you dry (another good reason to stuff a few garbage bags in your survival suit), but if it is not available, you can use bark, large leaves, or layers of branches.

Construct a door that is weatherproof, insulated, and easily pulled into place when you are in the shelter. Make it easy to remove, too, for quick escape when you need to signal rescuers.

Check the quality of your shelter by looking inside. Every place you see light needs to be filled. While small holes may seem insignificant, they will let in the wind, rain, and snow, while letting your body heat out. If your shelter is too big, pile more branches on the bed, the sides, or on top of you.

A good shelter will help keep you alive until rescuers can find you. Will they be able to see you and will they know you need help?

Signals

Two men were stranded on Chirikof Island. They felt certain they could signal one of the many passing boats, so they built a fire to attract attention. Fishermen saw the big blaze and talked about it, but didn't associate it with an emergency.

After seeing one too many boats go by, the men scoured the beaches for crab pot buoys which they mounded into several huge piles. They then built three massive fires and stood on the beach waving makeshift flags.

People began to think something might be wrong and someone notified the Coast Guard. The men finally were picked up. Their rescue took longer than they'd hoped because their first fire had not made it clear they wanted assistance. One bonfire looked like a beach party or a hunting camp to passing boats. Three fires in a line, mounds of orange buoys, and waving flags said, "HELP!"

To be most effective, emergency signals must attract attention and say, "HELP!"

The best way to attract attention is to make your signals look different from your surroundings. Use square angles, movement, contrast, and noise to your advantage.

Make sure you have signals that can be seen from the air and others that will be visible from the water. The U.S. Coast Guard recommends that each letter in your SOS signal be at least 18 feet long and three feet wide to be visible from the air. Place seashore signals well above the high tides to avoid rebuilding.

Signals in groups of three indicate distress, so if you build fires, build three, or hang three space blankets side-by-side. Other signals such as flares and dye markers relay the message for help when used singly.

Use both passive and active signals, and build several of each. Passive signals—like EPIRBs (Emergency Position Indicator Radio Beacons), flags, SOS, or three buoys hung in a row—work by themselves. Active signals are ones you operate. Flash signal mirrors at the horizon whenever you have a chance; you may contact a vessel you cannot see with your naked eye. Save some of your active signals such as flares for when rescuers are in sight.

On shore, signals are limited only by your imagination and available materials.

Make sure signals attract attention and convey the need for help.

Water

Water is critical for survival. We all need to drink 2 to 4 quarts of uncontaminated water every day for normal body function. Most of us meet our daily water requirement by drinking coffee, tea, milk, water, juice, and other beverages, and by eating foods high in water content such as soups and berries.

Without adequate water, most people will die within a week. Before they die, they will be severely incapacitated by the following effects of dehydration:

- Being thirsty and craving cold, wet foods
- Dark urine or a burning sensation while urinating
- Headache
- Depression or dull mental state
- Lack of energy
- Chapped lips and parched skin
- Nausea
- Constipation

(Some of these signs and symptoms of dehydration can be caused by other factors, too.)

Drinking urine, sea water, or contaminated water can make you sick and dehydrated from vomiting and diarrhea. **Do not** drink them. Untreated water can be contaminated by several disease-causing organisms, some of them lethal.

One organism of particular concern in Alaska is giardia, a parasite invisible to the naked eye. Giardia is found in untreated water throughout much of the world. Although many animals spread the disease through their contaminated feces, in Alaska beavers are often implicated as carriers, and the disease is sometimes called beaver fever.

Signs and symptoms of giardia usually occur 10 to 14 days after drinking contaminated water. They include severe abdominal cramps, diarrhea, vomiting, a bloated stomach, and fatigue. Because most people need prescription medication to avoid recurring bouts of giardia, it is wisest to avoid the illness altogether. Don't let it complicate your survival situation.

The best way to prevent all water-borne diseases is to drink treated water. Although there are chemical and mechanical ways to treat water, the most practical method in an emergency is to boil the water briskly for 20 minutes. One minute of brisk boiling generally kills giardia, but the extra 19 minutes are needed to kill other organisms. Boil melted ice and snow, too—freezing does not kill the giardia cyst.

Some chemical treatments are available to purify water, but they are not 100 percent effective when the water is cold or cloudy. If you have no way to boil untreated water, get it from the safest available source. Rainwater is generally safer than water from a creek, river, stream, or muskeg. A plastic bag works well as a rainwater collector.

Water is essential for digestion, so don't eat if you don't have water to drink. (Berries and other foods high in water are an exception to this.) When you are low on water, conserve your own by avoiding excess motion and sweating.

Food

Food is a wonderful thing. It allows the body to repair itself and to resist infection, provides us with energy and body heat, keeps our brain functioning, and helps ward off depression. We socialize over food, sometimes plan our day around it, and spend an incredible amount of time thinking about it.

Food is important, but most people can survive for a month or more with little or no food. That is why food is number six in the Seven Steps to Survival.

Native cultures have flourished for centuries on natural foods found along the North Pacific coast. Food can often be found in abundance if you know what is edible and where to look.

Concentrate your food gathering efforts on plants, berries, seaweeds, animals in the intertidal zone, and fish. Stalking large game often expends more energy than it yields.

Learn to identify edible plants and animals. This knowledge can help you in an emergency and can enhance your daily diet at home. Consult the resources listed in the resources section in this book, or take an edible foods class for more information on edible plants and animals.

Many edible foods are available along coastal Alaska.

Paralytic Shellfish Poisoning

The Russians and Aleuts in Alexander Baranof's 1799 otter pelt gathering trip learned about Paralytic Shellfish Poisoning (PSP) the hard way. Feasting on mussels in a bay near Sitka, many were overcome by tingling and burning lips, gums, tongues, arms, and legs; shortness of breath; lack of coordination; and a choking sensation. Nearly 150 died from respiratory failure caused by PSP. That bay is now named Poison Cove.

PSP is caused by poisons in tiny organisms—called dinoflagellates—often found in North Pacific waters. Clams, mussels, geoducks, oysters, snails, scallops, and barnacles are all potential carriers of this potent neurotoxin. Contrary to what many people believe, it is **not** safe to eat these shellfish in months whose names contain the letter R.

"The dinoflagellate responsible for the toxin is of the genus *Gonyaulax,* and the optimum temperature for its growth is 8°C [46°F]. This is also the mean sea temperature for Southeast Alaska, making that part of the coast an excellent place for the toxin to flourish. The colder the sea temperature, the less productive the environment for toxin development. By the time you get to the Bering Strait, the guilty dinoflagellates are present in only very low numbers. This explains why clams have been . . . a mainstay of the local Bering Strait people's diet without the incidence of PSP found in other parts of the state.

". . . The time it takes clams and other bivalves to pass the toxins through their systems [can be] many years, not months. . . ." (Alaska Marine Safety Education Association *Marine Safety Instructor Training Manual.*)

Some Foods to Avoid

- Hairy triton (a greenish snail with brown, fuzzy hairs).
- Clams, cockles, mussels, scallops, oysters, geoducks, moon snails, and barnacles (all may be tainted with the toxin that causes Paralytic Shellfish Poisoning).
- Sculpin eggs (poisonous).
- Raw fish (cook it first to kill parasites).
- Starfish, coral, sea anemone, jellyfish, sponge, nudibranch, and sand dollar (all use toxins to kill their prey).
- Baneberry roots and berries, false hellebore, vetch, wild sweetpea, Nootka lupine, death camas, and poison water hemlock (which looks very similar to wild celery).

- Some mushrooms (mushrooms have little nutritional value, so it is best to avoid them unless you can distinguish between the poisonous and edible ones).

- *Desmarestia ligulata* (a brown, branch-like seaweed that turns green when taken out of the water and smells like sulfur when it is handled).

Play

You have recognized the emergency, done an inventory, built a shelter, put up signals, discovered a source of water to boil, and have even found some food. Now what? Play!

Play means staying mentally and physically busy, improving your shelter, signals, water catchment and cooking arrangements, sharing stories and jokes, and emphasizing the positive.

Continue to act like a survivor, not like a victim. Play will keep your spirits up and strengthen your will to survive. Review the Seven Steps every time your situation changes.

Fire

Are you wondering why fire is not one of the Seven Steps? Although it can be an important part of your shelter, signals, water, food, and play, and can provide a psychological boost, it is not absolutely necessary for survival. There have been cases of people surviving in coastal Alaska in the winter for **weeks** without a fire, even though they were wet when they began their survival experience. One of the important factors in their survival was finding and making shelter.

There is a danger in over-emphasizing the use of fire, especially if you are in a wet environment and are not used to building a fire in these conditions. To be an effective heat source, outdoor fires need both abundant fuel and fairly constant attention. People have tried and failed to build fires, and then died because they neglected

TINDER
Dry—the size of grass and pine needles

KINDLING
Dry—twigs and small branches up to the size of your little finger

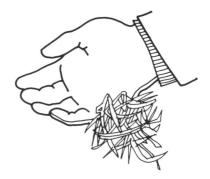

FUEL
Larger than finger in size

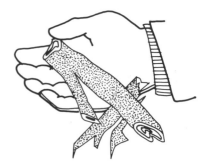

Tinder, kindling, and fuel are needed to start a fire.

to build a shelter and signals.

Fire can make your situation easier, but you **can** survive without it. If you do choose to build a fire, there are several principles to remember:

- Fires need fuel, heat, oxygen, and the chemical chain reaction between these three elements to burn.

- Building a fire takes patience, tinder, kindling, and fuel.

Don't Give Up

It is described as one of the most incredible tales of marine survival. Although not a fishing trip, the lessons learned from Ernest Shackleton's 1914 expedition to the South Pole are relevant for today's North Pacific fishermen.

On December 5, 1914, Shackleton and 27 men left South Georgia Island aboard the *Endurance* and headed for Antarctica. By January 18, 1915, their ship was trapped in ice in the Weddell Sea. The men and the ice-bound ship drifted north for ten months until *Endurance* was crushed and sank on November 21, 1915.

"For [the next] five months the whole ship's party . . . drifted north on a huge ice . . . floe that shattered and shrank as time passed. . . . Cracks opened up under tents [and] camps had to be changed with desperate speed. . . .

"Shackleton displayed his superb leadership during this very trying period—keeping everyone busy, making alternative preparations for any eventualities, and maintaining morale with jokes, entertainments and special treats. . . . One evening he even cheerfully discussed taking an expedition to Alaska when the present one was finished. . . .

"During January, February and March the surface of the ice floes was soft and mushy and living was continuously wet and uncomfortable, with an increased sense of insecurity and appalling boredom. . . .

"On April 9 the pack ice separated and the boats were quickly launched. Then followed seven days of constant crises and danger as they jostled through the rest of the pack ice and finally out into the wild open seas to an ultimate landing on April 15 on desolate Elephant Island. . . . Nobody had any way of knowing they were there." (*Shackleton's Boat Journey*, F.A. Worsley, W.W. Norton and Company, Inc., New York, 1977, pp. 25-27.)

Nine days later Shackleton and five of his men launched a 22-

foot open boat in an attempt to reach help on South Georgia Island. They battled monstrous seas and poor weather for more than 800 miles until they finally landed on the southwest coast of the island— a formidable navigation feat. With three of his crew incapacitated, Shackleton and two others set off across glaciers and snowy passes in an effort to reach the whaling station on the other side of the island.

Meanwhile the remaining 22 men on Elephant Island faced isolation, cold, and the anxiety of wondering if they would be rescued. They had no way of knowing that Shackleton and his men had reached help until a rescue vessel came into sight on August 30, 1916, more than 19 months after *Endurance* had become trapped in the ice. Amazingly, all 28 men survived the ordeal.

Do not underestimate the will to live. People often survive what they *believe* they can survive, some in seemingly unlivable circumstances.

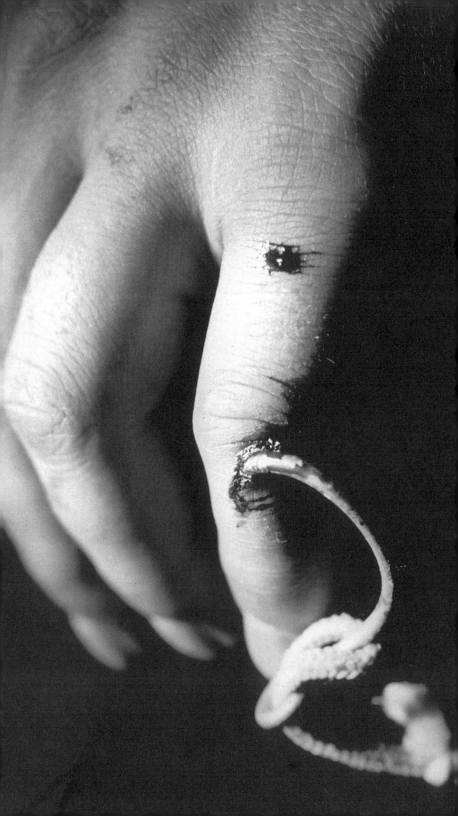

First Aid Afloat

Even the most careful skipper can have an unexpected onboard medical emergency. Suppose rough seas throw a crew member hard enough to break his leg. Or one of the newer crew gets his arm caught in the winch before it can be stopped. He bleeds a lot, and he looks like he will faint. Do you know what to do?

Do This First:

Before treating anyone, assess the situation and decide what to do to protect yourself. Watch for swinging booms, moving pots, hot liquids, fire, or other dangers that can harm you.

Then, answer these five questions: What caused the accident? Does the person respond to you? Does he have an airway and is he breathing? Does he have a pulse? Is he bleeding a lot? **If he doesn't have an airway, isn't breathing, doesn't have a pulse, or is bleeding a lot, you need to act fast!**

1. What caused the accident?

Look around for clues, and ask others what happened. Proper assessment of the situation can tell you a great deal about what might be wrong with the person. Someone who gets hit in the chest with a crab pot or other heavy object might have broken ribs, punctured lungs, breathing problems, or heart damage.

2. Does the person respond to you?

If the person isn't making any noise, try to get a response by gently tapping him and asking, "Are you okay?" If he doesn't respond, he needs help!

Yell for help if people are nearby. Contact help on the radio.

Encourage the person by being positive in your speech and actions. Unconscious people can often hear what you are saying.

3. Does the person have an airway, and is he breathing?

If he is talking or screaming, he has an airway and is breathing.

If you are not sure if he is breathing, look, listen, and feel for 3 to 5 seconds. You can usually do this without moving him. If he is breathing, find out if he has a pulse as in step 4. If he is not breathing, you must breathe for him as follows:

If he isn't on his back, carefully turn him over, keeping his neck and back in line.

Turn the accident victim over so he is on his back.

Kneel beside his head. If his neck or spine could be injured, use the jaw-thrust maneuver to open his airway. Put your thumbs on his cheekbones and your fingers under the corners of his jawbone and—without tipping his neck—lift his jaw up to get his tongue off the back of his throat. If you are sure he doesn't have a neck or spine injury, use the head-tilt/chin-lift maneuver to open his airway.

Check again for 3 to 5 seconds to see if he is breathing. If he has started breathing, find out if he has a pulse as in step 4. If he is **not** yet breathing, breathe for him. Keep his head tipped, pinch his nose shut, take a deep breath, and breathe into his mouth until his chest rises, then immediately give him a second breath.

*Use the jaw-thrust maneuver to open the airway
of a victim whose neck or spine may be injured.*

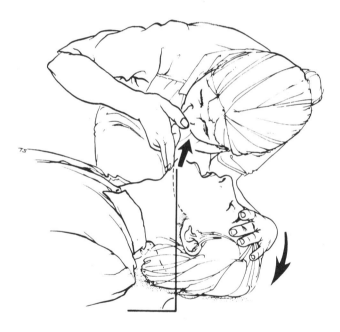

*If you are sure the person does not have a neck or spine injury,
use the head-tilt/chin-lift method to open the airway.*

If his chest does not rise, reposition his airway, and check to make sure his nose is pinched closed and you have a tight seal around his mouth. Try to give the two breaths again. If air still won't go in, turn to the choking section on page 126.

4. Does he have a pulse?

Check for a carotid pulse for 5 to 10 seconds. Check the pulse of drowning and hypothermia victims for up to 45 seconds. This pulse is in the neck, in the hollow between the windpipe and the large neck muscles.

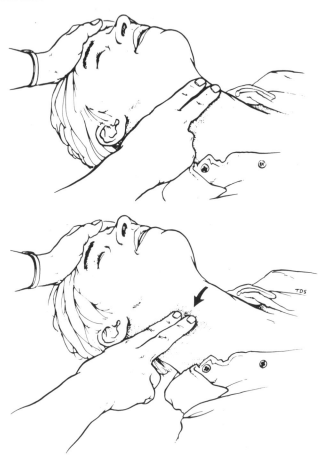

Locate and check the carotid pulse.

If he has a pulse and is breathing, loosen any restrictive clothing, and then check for bleeding as in step 5.

If he has a pulse, but is **not** breathing, breathe for him once every 5 seconds until he breathes on his own, someone else takes over, or a doctor tells you to stop. Recheck his pulse every minute or so.

If he does not have a pulse, do CPR (Cardiopulmonary Resuscitation).

CPR

To find the proper hand position, trace along the bottom of his ribs until you feel his breastbone, place the heel of your other hand two finger-widths up from the tip of his breastbone, then lay your first hand on top. Make sure your hands are not on his ribs.

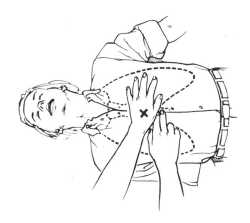

Locate the proper hand position for CPR.

Compress his chest 1½ to 2 inches, then release the pressure. Do 15 of these compressions in 10 to 12 seconds.

Open his airway and give him 2 full breaths.

Repeat the set of 15 compressions and 2 breaths until you have done a total of 4 sets. (Four sets should take about one minute.)

Check his pulse for 3 to 5 seconds. If he has a pulse, make sure he is breathing. If he is, place him on his right side (recovery position).

If you have not contacted the Coast Guard or a physician, do so now.

If he does **not** have a pulse, continue your cycle of 15 compressions and 2 breaths, rechecking his pulse every few minutes. Continue CPR until the person breathes on his own and has a pulse, someone else takes over, you are too tired to continue, or a doctor tells you to stop. It may take you hours to revive a drowning victim.

One-Rescuer CPR Standards

Age of patient	Pulse checked	For compressions use	Depth of compressions	Number of compressions to ventilations
8 and over	Carotid (neck)	Two hands	1½" to 2"	15:2
1-8 years	Carotid (neck)	One hand	1" to 1½"	5:1
Less than 1 year	Brachial (upper arm)	Two fingers	½" to 1"	5:1

CPR is very tiring. You may become so exhausted that you have to stop.

Taking a CPR course is the best way to learn, but try to do CPR in an emergency even if you've never done it before.

5. Is he bleeding a lot?

Look and feel for bleeding as you quickly run your hands under him.

Immediately place **direct pressure** on all severely bleeding wounds. Use a clean bandage or cloth. Use slightly less pressure for head wounds so you don't push broken bones into his brain. While maintaining direct pressure, **elevate** the injured part if it is an arm or a leg. Maintain direct pressure and elevation until the bleeding stops.

If the bleeding is not controlled in 30 seconds, check to make sure you are applying pressure to the bleeding site, then resume direct pressure and elevation.

If direct pressure and elevation do not control the bleeding on an arm or leg within one minute, use a **pressure point**—a place where an artery lies near the skin and on top of a bone—to slow blood flow to the wound. Use the brachial (upper arm) pressure point for bleeding arms, the femoral (in groin area) pressure point for bleeding legs.

Continue to use direct pressure, elevation, and pressure points until the bleeding stops.

Maintain direct pressure and elevation until the bleeding stops.

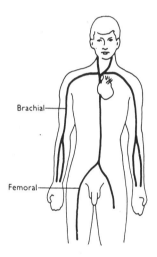

If necessary, press on the brachial pressure points when the arms are bleeding, and the femoral pressure points when the legs are bleeding.

Direct pressure and elevation will control most bleeding. **Do not** use tourniquets to control bleeding. They often do more harm than good, and the victim usually loses the limb.

If there is an **amputation**, control bleeding from the stump by direct pressure, elevation, and pressure points. Find the amputated part, keep it clean, dry, and cool (not frozen), and transport it with the person.

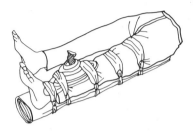

Stabilize impaled objects.

If there is a knife, hook, or other object stuck in the wound **do not** remove it. Stabilize the object by bandaging around it.

Treat the person for shock by placing a blanket over and under him, and elevating his legs 12". If this position causes more pain, lower his legs. Act calmly, and do not give him anything to drink or eat.

Look for and treat other injuries.

Contact the Coast Guard or a physician to determine further treatment.

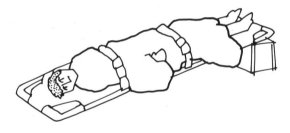

Treat shock by elevating legs 12".

Treating Other Injuries

Choking

People who are choking often grab their throats. Some will cough. **Do not** interfere with coughing. Others will be unable to cough or speak.

If you suspect someone is choking, ask him if he can talk. If he can, leave him alone unless his condition worsens.

If he cannot talk, stand behind him and place your fist on his stomach between his belly button and ribs. Put your other hand on top of the first hand and—without touching his ribs—swiftly pull

*Hand position for abdominal thrusts
on a conscious choking victim.*

both in and up. Repeat this motion until the object is out and the person can breathe again, or he becomes unconscious.

If you cannot get your hands around the person's stomach, or if she is pregnant, do the same maneuver on the chest, making sure you are in the **middle** of the breast bone.

If the person becomes unconscious, lay him on his back, open his mouth with the tongue-jaw lift, and sweep out any foreign objects. Then, open his airway with the head-tilt/chin-lift maneuver and attempt to breathe for him. If his chest rises, breathe for him every 5 seconds (checking his pulse every minute) until he breathes on

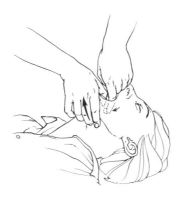

*Use the tongue-jaw lift on unconscious choking
victims, then sweep out foreign objects.*

his own, someone relieves you, you are too exhausted to continue, or a physician tells you to stop. If he does not have a pulse, do CPR (see page 123). If his chest does not rise, continue to the next step.

Straddle his legs and perform 5 abdominal thrusts.

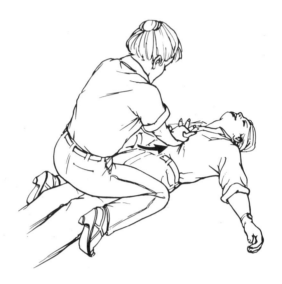

*Position for abdominal thrusts on an
unconscious choking victim.*

Then, repeat the finger sweep, breathing attempts, and abdominal thrust sequence until you are successful at getting in breaths, someone takes over, or a physician tells you to stop. Once breaths go in, you may need to do CPR.

Contact the Coast Guard or a physician if you have not done so.

Drowning

People who drown look cold, blue, and rigid. They are not breathing, don't have a pulse, and their pupils are big. However, it is sometimes possible to revive a drowned person if he is rescued from the water in time and CPR is started immediately.

Researchers believe that when a person drowns in cold water (water less than 70°F), the mammalian diving reflex extends his chance of living. This reflex, named after a similar response in sea mammals such as porpoises, whales, seals, and sea lions, causes the heart to slow and reduces circulation to the extremities. This allows the main blood flow to concentrate in the brain, heart, lungs, and kidneys, helping to preserve these delicate organs.

People who have been submerged under cold water less than one hour are most likely to be revived. CPR is not advised if the person is known to have been under water for more than one hour. If you are not sure how long the person has been under water, do CPR. When someone has drowned, get him out of the water as soon as possible, keeping him in a horizontal position if this does not delay rescue. Be extra careful of his back and neck if a spinal injury is suspected.

Start CPR as soon as he is out of the water (page 123). Do not do any special maneuvers to remove water from his lungs.

Contact the Coast Guard or a physician for further advice.

Heart Attack

Heart attacks occur when the heart muscle does not get enough oxygen and some heart muscle dies. Heart attacks can, but do not always, cause the heart to stop beating.

One of the most common symptoms of heart attacks is chest pain, often described as a squeezing sensation or "like someone is standing on my chest." This pain, however, is not always present. Some people have pain in the jaw or arms, especially the left arm. They may be sweaty, have pale or bluish skin, be short of breath, vomit, or feel nauseous, faint, or dizzy. Many patients deny they are having a heart attack, while others may feel sure they are going to die.

If you suspect that someone is having a heart attack, follow the five "Do this first" steps beginning on page 119, then return to this page for further treatment instructions.

Allow the person to sit or lie down so he is comfortable. Loosen tight clothing, reassure him, and try to be calm.

Give him oxygen, if possible.

Contact the Coast Guard or a doctor for further instructions, and make sure you continue to monitor his airway, breathing, and pulse.

Chest Injuries

Chest injuries can be caused by a blow or wound to the chest and can be life-threatening. If a lung is punctured, the person may cough up frothy, bright red blood. This can happen with or without a visible wound.

When a person has a chest injury, follow the five "Do this first" steps beginning on page 119, then return to this page.

If the person has a gunshot wound, look for and bandage the **exit** wound (where the bullet went out). There may not be an exit wound—some bullets stay inside the patient.

If a wound is visible where the bullet went in, seal it immediately with a piece of plastic, tinfoil, or other airtight material. A bulky bandage or clean clothing can also be used, although airtight material works best. Tape the plastic or tinfoil to the chest on three sides to allow air to escape from, but not enter, the chest cavity.

Unseal the wound immediately if the person's breathing gets worse.

Check for and treat other injuries, then contact the Coast Guard or a doctor for futher instructions. Maintain the person's airway, breathing, and circulation.

Abdominal Injuries

Follow the five "Do this first" steps beginning on page 119, then return to this page for further instructions.

Bandage wounds with a clean, dry dressing or cloth. If the person's intestines are hanging out, do not push them back in; this can cause a serious infection. Instead, cover the intestines with a clean plastic bag, or clean dressing or cloth moistened with clean water.

Allow him to lie with his knees drawn up if he prefers that position.

Check for and treat other injuries, then contact a physician or the Coast Guard for further advice.

Burns

First degree burns are red; second degree burns are red, blistered, and quite painful; and third degree burns often look charred or leather-like. Feeling is lost in third degree burns, although the area around them may be quite painful.

Before approaching a person who has an electrical burn, make sure he is not in contact with the electricity, and be sure the power is off.

Stop the burning without burning yourself. If someone's clothes are on fire, stop him, have him drop to the ground, and roll. If you have a wool blanket or article of clothing, roll him in it.

Follow the five "Do this first" steps beginning on page 119, then return to this page for further advice. Be especially concerned about people who have been burned on the face, were in a smoky, closed, burning area, or who suffered an electrical burn.

If it has been less than 15 minutes since the burn, and if the burn is first or second degree with unbroken skin, cool the burn by putting it in clean, cool water. Or put cloths dipped in clean, cool water on the burn. Do this until the pain lessens.

Use a glove to brush off dry chemicals such as lime, then rinse the area with clean water for at least 10 minutes.

If the chemical is in the eyes, hold the eyes open and rinse them gently for at least 10 minutes with clean, warm water. Flush alkali burns (ammonia, bleach, strong detergents, lye, etc.) for at least one hour.

Rinse chemical-contaminated eyes carefully for at least 10 minutes with clean, warm water.

Bandage the burned area (including entrance and exit wounds from electrical burns) with clean, dry bandages or cloths. Also bandage between burned fingers and toes, but do not put anything else on a burn unless directed to do so by a physician.

Check for and treat other injuries, then contact the Coast Guard or a physician to determine further treatment. Be prepared to describe the person's condition, how the burn happened, what it looks like, and how much of the body it covers.

Carbon Monoxide Poisoning

Carbon monoxide poisoning—and death—may occur if a vessel has a leaky exhaust system or if the wind is at the stern of the vessel. Air circulation around smaller vessels with cabins may cause the exhaust to cycle up toward the stern. Make sure you keep living and working spaces well ventilated, and do not use charcoal heaters inside boats.

Buy some carbon monoxide detectors for your vessel. They are relatively inexpensive and can help prevent this poisoning.

Signs and symptoms of carbon monoxide poisoning include headache, drowsiness, nausea, bright red or bluish skin, and unconsciousness. Some of these symptoms are the same as seasickness. When in doubt, suspect carbon monoxide poisoning and act accordingly. Rescue the person from the area and get him into fresh air, but make sure you are not overcome by the fumes yourself!

Follow the five "Do this first" steps beginning on page 119, then return to this page for further instructions.

Give the person oxygen, if possible. Check for and treat other injuries, then contact the Coast Guard or a physician for further advice.

Hypothermia

Hypothermia occurs when the body's core temperature drops. Submersion in cold water is a major cause of hypothermia because water conducts heat away from the body 25 times faster than air of the same temperature. Hypothermia can also result from a combination of wind and cool or cold temperatures, wet clothing, or clothing that is not suitable for the weather.

Although hypothermia can easily occur when air temperatures are above freezing, it can be prevented by using good judgment, wearing layered clothing to stay warm but not sweaty, putting on rain gear before getting wet, and avoiding being immersed in cold water. It helps to remember that 50 percent of your body's heat is lost through your head and neck. Other high heat loss areas are your armpits, chest, and groin.

Signs and symptoms of hypothermia include uncontrolled shivering (although some people do not shiver), confusion, poor coordination, slurred or slow speech, poor judgment, and drowsiness. A weak or irregular pulse, slower and shallower breathing, dilated (big) pupils, and unconsciousness can also occur. It is sometimes difficult to detect hypothermia because the affected person may not know, or may deny, that he is having a problem. In addition, signs and symptoms may be confused with or complicated by alcohol.

If you suspect that someone has hypothermia, check the person's pulse for up to 45 seconds when doing the steps on page 119, then return to this page for further advice.

Treat the person gently. If he is breathing and has a pulse, carefully remove his wet clothes and cover him with dry coverings.

If you are less than 15 minutes from a medical facility, do not add heat. Check for and treat other injuries.

If you are more than 15 minutes from a medical facility, consider putting the person in a sleeping bag and providing skin-to-skin contact with a warm person. Keep the air temperature above 80°F if possible.

Do not use tourniquets or try to cool the arms and legs. If the patient is severely hypothermic do not put him in a shower or bath, and do not give him alcohol or coffee.

Make sure you check for and treat other injuries.

Give the person warm fluids only after uncontrolled shivering stops, he is alert enough to get a cup of hot drink to his mouth by himself without spilling it, and can swallow without choking.

Patients with the following signs and symptoms should be transported to a medical facility: lack of shivering despite being cold, slowed pulse or respirations, confusion or unconsciousness, or presence of an associated illness or injury.

Contact the Coast Guard or a doctor for further advice.

Fractures (Broken Bones)

Not all fractures are life-threatening, but they should be considered serious until proven otherwise by a physician. Fractures to the head, neck, or back are more dangerous than others.

If you think someone has a fracture, follow the five "Do this first" steps beginning on page 119, then return to this page.

Do not move suspected fractures before they are splinted unless the person's life is in danger. If you need to move him before splinting, support the fracture site and the joints above and below the fracture during the move.

Do not pull on the arm or leg, do not try to set the bone ends, and do not push projecting bone ends into the wound.

Splint the fracture using whatever materials are handy. Try to splint the fracture in the position it is in with a **padded** splint, making sure you immobilize the fracture site and the joints above and below the fracture.

If the person complains that the splint is too tight or if his fingers or toes turn blue when the limb is splinted, loosen but do not remove the splint.

Apply cold compresses to the fracture site to help reduce swelling.

Check for and treat other injuries, then contact the Coast Guard or a physician for further advice.

Back or Neck Injuries

If there is a chance the person has a back injury, treat him as if he does.

Moving people with back injuries is a very special skill. It should be practiced during training before it is done to an injured person. If you're a fisherman and you haven't taken a first aid course, you should sign up for one soon!

Signs and symptoms of back injuries may include a wound, pain at the site, numbness, tingling, lack of feeling, or inability to move the body below the injury site. **It is possible for a person to have a back injury and show none of these signs or symptoms.**

If you think someone might have a back or neck injury, follow the five "Do this first" steps beginning on page 119, then return to this page.

Look for and treat other injuries before moving the patient. Do not move him until you have enough people and the proper equipment; improper handling can paralyze a person for life.

If you must move the person, **do not bend his neck or spine.** Place him on a backboard, or something hard like a bin board, in the position he is in. Put a blanket on the board before you move him. Cover him and secure him to the board.

Contact the Coast Guard or a physician for further medical advice.

Head Injuries

Head injuries may be bloody, painful, or swollen, or the person may have bruising around his eyes or behind his ears. He may have breathing problems, his pupils may be unequal, he may have vision problems, or he may be having a seizure. He may be unconscious.

If a person has a head injury, assume that he also has an associated back or neck injury, especially if he has been knocked unconscious. Do not move the person unless you have to.

Follow the five "Do this first" steps beginning on page 119, and then return to this page.

Look for and treat other injuries, then contact a doctor or the Coast Guard for further advice.

Seizures

Seizures are caused by a massive electrical discharge in the brain, and they are often accompanied by convulsions (involuntary body movements). Seizures can be caused by epilepsy, old or recent head injuries, alcohol withdrawal, diabetic problems, poisoning, fevers, drugs, and low levels of oxygen in the brain.

When someone is having a seizure, **do not** try to restrain him, and **do not** put anything in his mouth. Protect him by clearing the area of sharp objects or items he might knock onto himself.

Contact the Coast Guard or a doctor immediately if a person has one seizure after another. This is a serious emergency.

When the convulsion is over, follow the five "Do this first" steps beginning on page 119, then return to this page.

Look for and treat other injuries, then contact the Coast Guard or a physician for further instructions.

Frostbite

Frostbite occurs when body tissue freezes. It can be prevented by wearing proper clothing and being prepared for the weather.

Frostbitten tissue usually looks pale or white. It is hard to the touch yet has no sensation.

To treat frostbite, follow the five "Do this first" steps beginning on page 119, then return to this page.

If the patient appears hypothermic, follow the treatment on page 132, then return to this page for thawing instructions.

Decide whether or not to thaw the frozen part. **Do not** thaw it if you cannot do it completely or if it has a chance of refreezing. If you do not thaw the part, protect it from thawing and further injury, and contact the Coast Guard or a physician for further advice. Do not rub the part or put ice or snow on it.

Thaw frostbitten parts in moving warm water (100° to 105°F) until normal color and sensation return. Then loosely bandage the part, placing bandages between thawed fingers and toes. Elevate the part, and prevent it from refreezing. Try to prevent the person from walking on thawed feet.

Contact the Coast Guard or a physician for further advice.

Paralytic Shellfish Poisoning

Paralytic Shellfish Poisoning (PSP) is caused by a poison produced by small organisms called dinoflagellates. Clams, mussels, oysters, snails, scallops, and barnacles ingest these organisms while feeding, and the poison is stored in their bodies. This toxin has been found in these seafoods **every** month of the year, and butter clams have been known to store the toxin for up to two years. One of the highest concentrations of PSP in the world is reported to be in the shellfish in southeast Alaska.

Some people have died after eating just one clam or mussel, others after eating many—each with a small amount of poison. You cannot tell whether the dinoflagellates are present by looking at the water with your naked eye. No simple, reliable test for PSP exists, and most beaches in Alaska are not tested. If you are not sure the seafood is toxin-free, avoid eating it if it is from an area with a high incidence of PSP.

Signs and symptoms of PSP most often occur within 10 to 30 minutes after eating affected seafood. Problems can include nausea, vomiting, diarrhea, abdominal pain, and tingling or burning lips, gums, tongue, face, neck, arms, legs, and toes. Later problems may include shortness of breath, dry mouth, a choking feeling, confused or slurred speech, and lack of coordination.

If you think someone has PSP, follow the five "Do this first" steps beginning on page 119, then return to this page.

If the person is conscious and alert, and can speak clearly, have him drink at least 2 glasses of water, each mixed with 3 tablespoons of activated charcoal.

Contact the Coast Guard or a physician for further advice.

Fishing-Related Injuries

Fish Hooks

When working around hooks, prevent getting hooked by thinking about your body placement.

If someone does get hooked, control bleeding with gentle direct pressure without removing the hook. If the hook is embedded in the eye, ear, nose, joint, bone, or other critical area, stabilize it where it is, and transport the person to a medical facility.

Remove fish hooks **only** if they are surface snags or you cannot get to a medical facility within 6 to 12 hours. If you are unsure whether or not to remove a hook, contact a physician or the Coast Guard for medical advice.

If you are going to remove the hook, wash the area and hook with an antiseptic solution such as Betadine™ and then hot, soapy water to lessen the chance of infection. Then numb the area with clean ice, and decide which hook removal method is best.

To remove **surface snags**, sterilize a razor blade or sharp knife with Betadine™ or heat, then cut through the skin to the barb and remove the hook.

If the hook is more deeply imbedded, use the **push and cut** method: Use needle-nose pliers to push the barb through the skin, cut the barb off with bolt cutters, and pull the rest of the hook out in the opposite direction.

The **flicker** method can also be used for small, deeply imbedded hooks. Do **not** use this method on circle hooks, as they collect too much tissue when removed this way. First, put the hooked body-part on a firm surface and hold the curve of the hook with needle-nose pliers. Then, use your finger to push down on the shank of the hook to disengage the barb, and quickly pull hard on the pliers.

After the hook is removed, wash the wound with Betadine™, and then hot, soapy water. Bandage the wound to reduce the chance of infection, and contact a physician for further instructions. The person may need a tetanus shot.

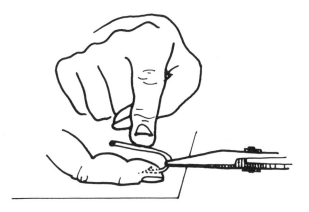

*Use the flicker method for removing
small, deeply imbedded fish hooks.*

Infections from Handling Fish

Infections from handling fish (sometimes called fish poisoning) can develop when bacteria from the fish enters your body through cuts, scrapes, or punctures.

Antibiotics such as Keflex™ or erythromycin are commonly prescribed for fish poisoning. Before the fishing season begins, ask your doctor for a prescription for an antibiotic to take along in case you're a long way from port.

Prevention includes trying to keep your hands and gloves clean and dry, changing or washing out your gloves each day, and washing your hands with Betadine™ and then hot, soapy water at least twice a day.

Swelling and redness at the wound site is common and can increase as the infection spreads. A fever or chills may also develop with a worsening infection.

Wash wounds with Betadine™, and then hot, soapy water as soon as they occur. Then dry and bandage the wounds.

If the wound looks infected, soak it for ½ hour in hot, soapy water (as hot as you can stand without burning yourself) at least three times a day. Then dry and bandage the wound.

Some doctors recommend wrapping a wet, room-temperature tea bag around the wound for 10 minutes several times a day. Use regular, not herbal, tea bags. Then dry and bandage the wound.

Contact your doctor or the Coast Guard if the infection gets worse or does not clear up in a few days.

In very severe cases, surgery may be necessary to drain the pus from the infection.

Fish Punctures

Spine sticks from northern sculpin, ratfish, short spine thornyhead, and a number of rockfish and other spiny fish can cause a serious infection or death if they are left untreated.

Signs and symptoms of fish punctures can include trouble breathing, a painful cut or puncture, swelling, nausea, vomiting, cramps, and paralysis. Extreme tenderness and a fever are signs of a spreading infection.

Treat all fish punctures as soon as they happen. Follow the five "Do this first" steps beginning on page 119, then return to this page.

Carefully pull the spine straight out, making sure you get all of it. Be gentle—the spine breaks like glass. Save the spine so a doctor can check it for broken remnants left in the body.

Northern Sculpin

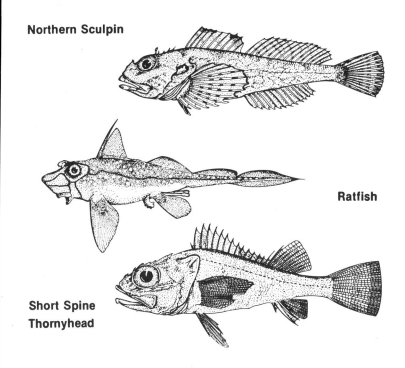

Ratfish

**Short Spine
Thornyhead**

*Spine punctures from spiny fish can cause
serious infection or death if they are not treated.*

Wash the wound with Betadine™, and then hot, soapy water.
Then soak the wound for at least 30 minutes in water as hot as you
can stand without burning yourself. Hand or dish soap, or
Betadine™ can be added to the water. Bandage the wound and
keep it clean and dry.

If swelling or redness occurs, soak the wound at least three
times a day in hot, soapy water until the infection clears.

Contact a doctor or the Coast Guard if the wound becomes extremely tender or you develop a fever.

Jellyfish Stings

Jellyfish stings occur when the stinging cells on the jellyfish's
tentacles touch your skin. For fishermen, this is most likely to occur
when pulling line, taking fish out of nets, or hauling nets.

You can prevent most stings by wearing rain gear, gloves, and
goggles, and by smearing petroleum jelly on your face before you

begin pulling nets. Make sure your hands are clean before you wipe or rub your eyes.

When the stinging cells hit your skin, they release a poison that can cause a temporary burning pain and a skin rash. In serious cases, the stings may cause difficulty breathing, shock, nausea, vomiting, or cramps.

To treat jellyfish stings, follow the five "Do this first" steps beginning on page 119, then return to this page.

If tentacles are in the eyes, rinse them with a clean saline solution (or clean water if saline is not available) until the burning stops.

If the tentacles are not in the eyes, pour vinegar on the site to help prevent the stinging cells from firing. Put a baking soda and water paste on the site for 15 minutes, then wash the paste off with saline solution or clean sea water. Put another baking soda and water paste on the site for 5 minutes, then gently scrape the paste off with a knife.

Give aspirin or other pain relief medication to help control pain.

In severe cases, contact the Coast Guard or a doctor for further advice.

Carpal-Tunnel Syndrome

Carpal-tunnel syndrome is a common affliction, especially among longliners and cannery workers. Tendons are tough, long cords that connect muscles to bone. Any abnormal strain on the tendons can cause chafing and swelling, which creates pressure on the median nerve as it passes through the tight carpal tunnel in the wrist.

You can prevent carpal-tunnel syndrome by doing arm, wrist, and hand exercises before the fishing season opens. Doing stretch exercises to limber up the wrist and fingers before work and after long closures is also helpful.

Compression of the median nerve can cause a weak grip, clumsiness, numbness, and burning pain (which is often worse at night). Neglected carpal-tunnel problems can lead to permanent damage.

Treatment includes resting the injured wrist, wearing a splint at night (and during the day if possible), and taking aspirin or ibuprofen before, during, and after heavy wrist use. Although some physicians think cortisone injections are effective, others believe that they can cause weakness in the affected hand.

Surgery may be necessary to open the carpal tunnel in severe cases.

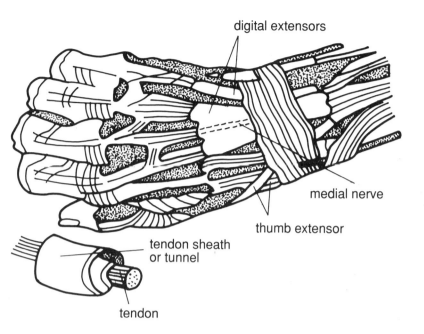

digital extensors

medial nerve

thumb extensor

tendon sheath
or tunnel

tendon

Medial nerve position, tendons, and tendon sheaths on the wrist.

Tendonitis

Tendonitis results when a tendon is overworked, and the tendon and the sheath surrounding it become swollen and painful. Fishermen are most likely to experience tendonitis in their wrists, fingers, thumbs, and elbows because of repetitive or hard movements such as pulling pots and lines, and baiting hooks for hours on end.

Many cases of tendonitis can be prevented by taking time to strengthen muscles and tendons in your upper body before the fishing season begins. During the season, try to use smooth rather than jerky motions, and chest and abdominal muscles instead of shoulder or arm muscles. Reduce wrist strain by cutting hard, frozen bait with an ulu, cleaver, or hatchet instead of a regular knife. Periodically switch hands when performing repetitive tasks, and hold lines with the thumb gripping on the same side as the fingers to minimize pressure on the hand.

Aspirin or other non-steroid, anti-inflammatory drugs can prevent the initial swelling, especially if taken before an opening. Take large doses only under a physician's direction.

One early sign of tendonitis is swelling (although it does not always occur). Another is squeaking or creaking at the injury site. This creaking can be felt by applying slight pressure at the site while moving the joint. Pain and stiffness in the palm or underside of the fingers or thumb are also common.

Tendonitis is aggravated by hitting or otherwise injuring the area, continuing to use the injured joint, and wearing constricting clothing and wrist bands.

Rest is usually required for tendonitis to totally heal. A few treatments may help fishermen who must continue working despite the pain: Reduce swelling by soaking the affected joint(s) in cold water or applying cold packs on the area to help reduce the swelling. Other people report that hot packs help. Use whichever works for you. Aspirin may help reduce the pain.

Tendonitis is not a permanent condition, although scar tissue may form on injured tendons in severe cases. A serious injury may take several months to return to normal or may require surgery. These are compelling reasons to take steps to prevent tendonitis from occurring.

Helicopter Evacuation

Evacuating patients by helicopter is a hazardous operation for both the patient and the aircraft crew. A small vessel, high masts, confined work area, high seas, gusty winds, and darkness compound the hazard. Evacuations will be attempted only in the event of a serious injury or illness.

Gather the following information. The Coast Guard needs it before they can decide how to help:

- Vessel name, call sign, position, course, and speed
- Nature and time of injury
- Patient's name, sex, age, and nationality
- Patient's pulse rate, breathing rate, blood pressure, and temperature (if known)
- Amount of blood loss, other significant symptoms, present medication or treatment being given
- Whether you need Coast Guard assistance
- Local wind direction and speed, sea state, and cloud cover

Contact the Coast Guard.

Preparations Prior to Helicopter Arrival

Maintain continuous radio guard on VHF channel 16, HF 2182 or 4125 KHz, or specified voice frequency. You may receive medical advice, positioning instructions, or be told to head for a rendezvous point. Advise the Coast Guard immediately if there is a change in any previously relayed information, especially changes in the patient's condition.

The air crew will discuss the most suitable hoist area with you. Pilots and crew generally prefer areas to the stern of the vessel with minimal obstructions.

Because rotor wash approaches 100 mph, it is important to secure loose gear, awnings, running rigging, and booms. Keep antennas up to maintain radio contact.

If the hoist is at night, light the pickup area and any obstructions, but **do not** shine any lights on the helicopter or use a flash camera—they can blind the pilot.

Because helicopters are noisy and voice communication will be almost impossible, it is important to prearrange a set of hand signals among the assisting crew.

Change course to permit the ship to ride as easily as possible with the relative wind 30 to 45 degrees off the port bow. Find the best speed to ease the ship's motion but maintain steerage way.

Place a PFD on the patient if his condition permits it, and put information about the patient's condition in his pocket or some secure place. Make sure it will not be blown away by the rotor wash.

Hoist Operations

If the patient's condition permits, move him close to the hoist area, but be alert to the dangers of rotor wash. If a litter is required, the Coast Guard will lower one.

If you do not have radio contact with the helicopter, signal "Come On" with your hand when you are ready for the hoist operation to begin. The helicopter will light the area at night.

To avoid static shock, allow the helicopter's basket or stretcher to touch the deck prior to handling it. If a trail line is dropped by the helicopter, use it to guide the basket or stretcher to the deck. It will not shock you. You may need to pull the rescue device to the vessel. Do not stand on or in front of the line and do **not** tie it off.

If it is necessary to take the litter away from the hoist point, unhook the hoist cable so the helicopter can haul it in. **Do not secure**

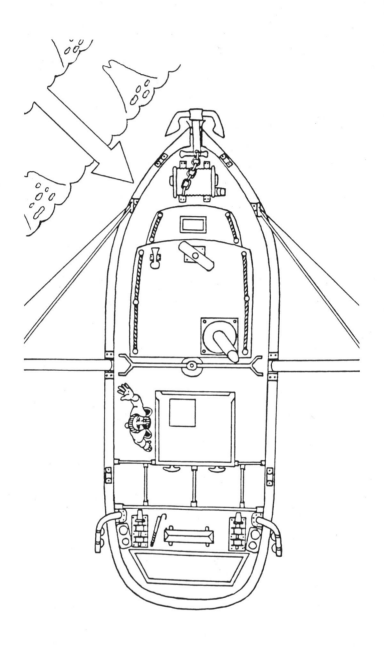

the cable to the vessel or attempt to move the stretcher without unhooking it.

If a basket is used, sit the patient in it, with his hands inside. If a stretcher is used, lay him in it face up and strap him in. If cable has been unhooked, signal the helicopter to lower the cable and hook up.

Signal the helicopter hoist operator when ready for the hoist: patient nods his head if he is able, and deck personnel give thumbs up. If the trail line is attached, use it to steady the stretcher or basket, keeping your feet clear of the line. Do not get between the device and the rail.

If time permits, the helicopter crew will retrieve the trail line.

Filing a Form

You may be required to file a form with the Coast Guard when an accident or injury occurs on board your vessel. Check the resources section in this book to determine when.

The information covered in this chapter will help during an emergency, but it is not intended to take the place of first aid training, which should be offered to all crew members.

First Aid Kit

Stock your first aid kit for the type of injuries you are most likely to encounter. The items below are for a basic kit; your physician may suggest including other items.

1" x 3" adhesive bandages

1" and 2" adhesive tape

4" x 4" sterile first aid dressings

2" x 10 yards gauze bandages

14" x 14" sterile dressings

Triangular bandages

Aspirin or ibuprofen

Activated charcoal

Tweezers

Betadine™

First aid kit contents, continued

> Needle-nose pliers (for removing hooks)
>
> Razor blades or a sharp knife (for removing fish hooks)
>
> Matches (for sterilizing razor blades or knives)
>
> Bolt cutters (for removing fish hooks)
>
> Prescription antibiotics and other needed prescription medications (Be sure your crew members are not allergic to them.)

First Aid Training Requirement

The regulations stemming from the Commercial Fishing Industry Vessel Safety Act of 1988 mandate first aid and CPR training for certain vessels. After September 1, 1993, documented vessels with more than two individuals on board must have at least one person certified in First Aid and at least one person certified in CPR. (The same person can hold both certifications.) Documented vessels with more than 16 persons on board must have at least two persons trained in First Aid and at least two persons trained in CPR. (The same two people can hold both certifications.)

Fire Fighting and Fire Prevention

After completing repairs, resupplying, and fire-fighting training in Seattle, the crabber headed to the Bering Sea. Near St. Lawrence Island, a fire was detected. The alarm was sounded, fire extinguishers were grabbed, and the crew approached the still relatively small blaze by staying low, just as they had been taught. Despite the early warning, they never got the fire under control because every extinguisher was empty. The crew ended up abandoning ship.

Seem unlikely? Unfortunately, an incident similar to this occurred in 1988. What sunk this crabber was a lack of usable fire extinguishers.

To successfully battle fires, every area of a vessel needs functioning portable fire extinguishers, crew members who are trained to operate them, and an automatic fire-fighting system (for larger vessels).

Fires

In order to burn, a fire must have fuel, heat, and oxygen—plus the chemical chain reaction. All fire prevention and fire suppression is aimed at separating the combination of these four components.

Anything can fuel a fire as long as there is enough heat to vaporize it into gases. The reason gasoline and other petroleum liquids are so dangerous is that they vaporize at low temperatures.

Fires are categorized into four classes—A, B, C, or D—depending on their fuel.

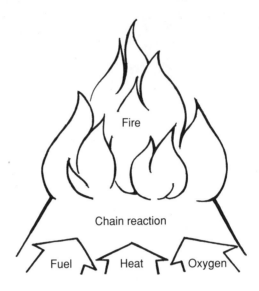

Fuel, heat, oxygen, and their chain reaction are needed for a fire to burn.

Types of Fires

- Class A fires are fueled by ordinary combustible materials such as wood, bedding, clothing, canvas, rope, and paper.
- Class B fires are fueled by flammable or combustible liquids such as gasoline, oil, paint, grease, etc., and flammable or combustible gases such as acetylene and propane.
- Class C fires are electrical fires.
- Class D fires are those caused by combustible metals. The only Class D metals aboard most fishing vessels are in flares.

Fires can be a combination of several classes. An engine room fire would most likely be a BC fire, but could easily be an ABC fire. The key to successful fire fighting is to choose the extinguishing agent that will put out the fire.

Fixed and Portable Extinguishers

Every fire-extinguishing agent, whether in a portable or a fixed system, puts out fires by eliminating fuel, heat, or oxygen, or by

Extinguishing Agent	How It Works
Water	Removes the heat source by cooling the fire.
Carbon Dioxide (CO_2)	Deprives the fire of oxygen.
Foam	Cools the fire and deprives it of oxygen.
Dry Chemical	Interferes with the chemical reaction. May cool, smother, and provide radiant shielding.
Halon	Interferes with the chemical reaction.
Dry Powder	Removes enough heat to bring the material below the ignition point.

breaking the chemical reaction between these components.

In order to snuff out a fire, an extinguishing agent needs to be properly directed. That's why trained crew members are a vital component of successful fire fighting. Even if you have an automatic system, you may still need to close vents and fuel lines, and shut down engines for the system to work properly.

Fixed Fire-Extinguishing Systems

A well-maintained fixed system using CO_2 or halon can detect and extinguish small fires before they become too large to fight. Unfortunately, CO_2 displaces oxygen, and halon can break down into toxins.

Portable Extinguishers

Portable fire extinguishers are classified and clearly marked by both a letter and number (except for class D fires, which have no number). The letters—A, B, C, or D—identify the class of fire the extinguisher will put out if it is used properly and the fire is not too large for the extinguisher. Some extinguishers work on more than one class of fire and will specify that on their label. For example, an extinguisher might be labeled BC and would be effective on class B, class C, or class BC fires.

The number indicates the size of the extinguisher. Although the Coast Guard uses the Roman numerals I, II, III, IV, or V to indicate the extinguisher size, the National Fire Protection Association (NFPA)

Types of Fires and Extinguishing Agents

Fire Type	Extinguishing Agents
A	Water works well. Multi-purpose dry chemical or ABC-rated extinguishers are also appropriate. Foam is excellent and penetrates better than water. Halon and carbon dioxide (CO_2) will work, but not as effectively.
B	Carbon dioxide (CO_2), foam, dry chemical, or halon are best. Water can be used as a fog or high volume spray on diesel fuel, but not on gasoline.
C	Carbon dioxide (CO_2) or halon works best. Dry chemical will also extinguish the fire but will ruin electronic equipment. CO_2 may also damage electronics by thermal shock. The extinguishing agent must be non-conducting.

uses Arabic numerals (2, 4, 5, etc.). In both cases, larger numbers indicate larger and heavier extinguishers. The Coast Guard system identifies the physical size of the extinguisher, while NFPA ratings indicate the amount of fire the extinguisher will put out. The two rating systems cannot be interchanged.

The above table indicates which extinguishing agent works best for class A, B, or C fires. Class D is not included, as these fires are rarely found on fishing vessels.

In order to put out a fire, you **must** use an extinguisher classified for that fire. Using a class A extinguisher on a class C fire will not put the fire out and could make it worse. Choose extinguishers for fires that are most likely to occur in a particular area. For ex-

Types of fires are designated by shapes and letters.

ample, extinguishers in engine rooms should be rated at least BC, as a fire in this location is likely to involve flammable liquids or gases (class B) and electrical equipment (class C).

Having the proper extinguisher is no guarantee that the fire will be put out. Crew members must be trained to use extinguishers effectively.

The Coast Guard puts out a fire on the F/V Ethel D in August 1983. (S. Anderson photo, Kodiak Daily Mirror)

Using Fixed Extinguishing Systems

For safety reasons, a manual activation device should be located outside the compartment containing the fixed fire-extinguishing system. Before a fixed system is discharged, everyone must be evacuated from the area.

In order for the halon or CO_2 system to work, the space where the fire is burning must be completely sealed. This means that hatches and doors must be closed, and ventilation and exhaust systems shut down or manually closed off. If the vessel is operating, both the fuel supply and electronic motors will need to be shut down to increase the likelihood of extinguishing the fire and reduce the chance of a reflash.

Because CO_2 and halon do not cool fires, spaces must be thoroughly cooled before they are ventilated. This can be done by cooling the exterior bulkheads and decks. Failure to adequately cool the area will cause the fire to flash back. Whenever possible, do not reopen a closed space where an automatic system has been triggered until adequate and professional fire-fighting resources are available.

Using Portable Extinguishers

The sequence for using portable extinguishers is:

1) **Pull** the pin.
2) **Aim** low.
3) **Squeeze** the trigger.
4) **Sweep** the base.

The pin on portable fire extinguishers must be **pulled** out before the extinguishers will work.

Aim the extinguisher at the base of the flame.

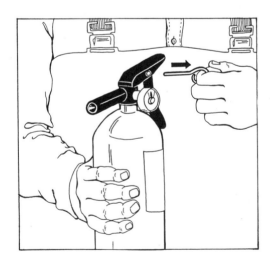

First, pull the pin on the extinguisher.

Aim low, squeeze the trigger, and sweep the base of the fire.

Keep your face away from the top of the extinguisher, and release the contents by **squeezing** the two handles together or by opening the valve. On a cartridge-operated, dry chemical extinguisher, the puncturing lever also must be hit with the palm of the hand. Do not hold directly onto the horn of a CO_2 extinguisher. It can get cold enough to cause frostbite.

As the extinguisher is discharging, **sweep** the base of the flame.

If the fire is electrical (class C), turn the electricity off and aim the extinguisher contents at the source of the fire. Work quickly and aim accurately—some CO_2 and halon extinguishers empty themselves in 8 to 10 seconds.

General Fire-Fighting Tips

- Be properly outfitted before attempting to fight a fire. If professional fire-fighting clothing is not available, wear wool clothing underneath rain gear (the rain gear acts as a vapor barrier against steam), and wear leather gloves—not synthetic or rubber. Do not wear cotton, polypropylene, or synthetic clothing; they ignite at low temperatures and will cause severe burns. In enclosed spaces, use a self-contained breathing apparatus (SCBA) with a safety line.

- **Always** keep your escape route open and clear. To maintain an uninterrupted flow of agent to the fire, have backup extinguishers and crew members immediately available.

- Although you must be fairly close to the fire to successfully fight it with portable extinguishers, do not approach the fire too quickly. Be familiar with the range of your extinguishers so you don't have to go closer than necessary.

- If you need to retreat, back away and keep your eyes on the fire. **Never** turn your back on a fire.

- Halon and CO_2 tend to blow away in windy conditions, so keep the wind at your back. This may mean turning the vessel.

- Fire fighting is dangerous business. Fires and decomposing halon produce poisonous gases, and both halon and CO_2 displace air. Because the vapors are invisible, you cannot determine their level of concentration by the amount of smoke in the air. When halon decomposes, it produces a sharp, acrid smell. This should be a signal to leave the area immediately. When using these extinguishing agents or fighting a fire in an enclosed area, you must either exit the area quickly or wear self-contained breathing apparatus.

Breaking the Fire Triangle and Stopping the Chain Reaction

1) Shut off engine(s), and fuel and gas lines that are feeding the fire.
2) Deprive the fire of oxygen by closing doors and hatches, and closing off ventilation systems.
3) Use the proper fire extinguisher for the class of fire.
4) Use your extinguishing agent wisely—it may only last 8 to 10 seconds.
5) Cool combustible materials before they ignite, to slow the fire's spread. This is especially important in metal boats—which conduct heat well.

Specific Tips for Classes of Fires

Class A fires will re-ignite if they are not totally cooled or covered with the extinguishing agent. Be very cautious working around burning fiberglass laminates, epoxies, and urethane insulating foam. They give off extremely toxic vapors.

Class B fires must be smothered or blanketed with the proper extinguishing agent. Be extra careful not to scatter the fuel while fighting these fires.

Shut the electricity off before attempting to extinguish class C fires.

Fire-Fighting Steps

Now that you are familiar with fire, fire extinguishers, and some basic fire-fighting tips, it's time to put it all together into a strategy. When a fire is detected on board, follow these five steps (some of these steps may occur simultaneously):

1) **Size up the emergency, notify the Coast Guard.** When a fire is detected, sound the alarm and get information on its type, location, and size. Notify the Coast Guard immediately of your problem and location. They can advise you on fire-fighting procedures and transport additional equipment to you.

   The fire-fighting method will depend on the vessel's arrangement, the location of the fire, and the available equipment. Every vessel should have a plan for fighting fires in all spaces.

2) **Rescue trapped people.** Check to make sure a crew member is really trapped before a rescue is attempted. Wear protective clothing and special breathing gear, stay low to avoid as much smoke and heat as possible, and **always** have a backup. You may need to extinguish the fire as you do the rescue.

3) **Confine the fire to its present size and location.** This is the time to shut doors and hatches, shut down engines, close off ventilation and exhaust systems, and turn off electricity and fuel lines in the fire area. Remember to check the fire's boundaries on all sides, bottom, and top.

4) **Extinguish the fire with the least damage to people and property.** A coordinated, trained crew will do steps three and four simultaneously, causing the least damage to the crew and contents. However, it is better to extinguish the fire with some damage, rather than to try to save the catch and lose everything.

   Make preparations to abandon ship while fire-fighting operations are taking place. Assign one crew member to prepare life rafts, etc. Abandon ship **only** if it is more dangerous to be on board than in the water. If a fire seems out of control, consider abandoning the vessel into a life raft, leaving the raft's painter line attached to the vessel until you are sure it should be cut.

5) **Overhaul.** This involves examining areas affected by the fire, cleaning up, and restoring machinery and equipment for operation. If water has been used to fight the fire, dewatering should begin immediately in order to maintain the vessel's stability.

Before opening closed areas where halon or CO_2 has been released, make sure they are sufficiently cooled. Where other extinguishing agents have been used, examine the fire area for hot spots or embers that need to be cooled or extinguished. Do not try to remove embers and burned material from the spaces. Instead, fill a drum or trash can with water, and immerse all involved materials before removal and disposal. Be prepared to fight new fires during the overhaul.

Training

The importance of training cannot be overemphasized, especially once at sea when the crew **are** the fire fighters. Away from port there is no fire department to call, and the Coast Guard is mandated to put out fires only when lives are in jeopardy.

All crew members should practice using the portable fire extinguishers and should know the basics of fighting fires, how to evacuate from all areas of the ship, and how to sound the alarm.

Fire Prevention

Preventing vessel fires involves common sense and taking time for maintenance checks.

If you can answer "yes" to the following questions, you are practicing good fire prevention. A "no" answer means you are flirting with fire danger.

Fire Extinguishers

___ Each month do you visually check all portable fire extinguishers and fixed systems? Do you tip all dry chemical extinguishers or hit them with a rubber hammer to make sure the chemical is loose?

TUPPER

___ Are your fire extinguishers and fixed systems checked by a certified person once a year?

___ When a fire extinguisher is used, do you have it refilled as soon as possible?

Engine Room

___ If your engine room is unattended, does it have a fixed automatic fire-extinguishing system?

___ Are there automatic engine and ventilation shutdown systems that operate before the extinguishing agent is discharged?

___ Are fuel connections tight?

___ Are remote fuel shutdown valves installed outside the engine room?

___ Are fuel and lubricating oil lines free of kinks? Are they replaced when brittle, cracked, or otherwise damaged? Are the lines arranged to prevent rubbing? Are they metal tubing, or if nonmetallic flexible hoses, are they rated for their intended service? Do they have spray shields at bends and connections to prevent atomization of fuel in case of a leak?

___ When exhaust lines lie near combustible material, are they insulated?

___ Are drip pans emptied frequently?

___ Is there a ventilated, covered, metal container for disposal of oily rags?

___ Are electrical motors regularly inspected, maintained, and replaced as needed?

___ Do you regularly inspect electrical wiring and hoses for cracking or damage, and replace them as needed with equipment approved for marine use?

___ Do the lights have vapor globes, and steel or plastic cages around the fixture?

___ Is the oil in your bilge kept to an absolute minimum?

___ Are charging batteries in a well-ventilated area? (These produce hydrogen, a highly explosive gas.)

Electrical System

___ Do you avoid overloading electrical outlets and motors?

___ Are loose, frayed, or worn electrical wires replaced, and short circuits promptly and properly repaired?

___ Are fuses and circuit breakers the proper size?

Accommodation Spaces

___ Are light bulbs covered and protected from contact with combustible materials such as gear and bedding?

___ Are space heaters positioned away from combustible materials?

___ Are there operational smoke detectors in each accommodation space? Are they tested each month?

___ Do crew members avoid storing oil-soaked clothes in crew lockers?

___ Have combustible curtains and carpets been replaced with noncombustible ones?

____ Do sleeping areas have an operable escape hatch overhead in case a fire blocks normal exit routes?

Galley

____ Are combustible materials stored a safe distance from the galley stove?

____ Is the stove turned off when it is unattended?

____ Are galley hoods, filters, and stacks cleaned regularly?

Smoking

____ Is smoking prohibited in bed and where combustible materials or flammable liquids or vapors are present?

____ Are cigarette butts and matches properly disposed of in ashtrays? (Cigarette butts thrown over the side can be blown back on board.)

____ Are ashtrays emptied into metal containers?

Combustible Materials

____ Are paints, thinners, solvents, and other combustible or flammable liquids properly stored in a designated locker or storeroom? Does the storage area have proper ventilation and sufficient fire-extinguishing equipment?

____ Do you store cardboard boxes, plywood, and other combustible materials away from heat sources?

____ Are all gas or hazardous materials cylinders stowed securely on deck in an upright position? (**Do not** store "heavier than air" gases above or near accommodation spaces.) Are the valves, pressure regulators, and pipes leading from these cylinders protected from damage?

Training

____ Does your crew know where the fire extinguishers are and how to use them?

____ Do you have regular fire drills?

Construction

___ Does your vessel include built-in fire endurance for bulkheads and decks?

___ Are noncombustible materials used for furniture, bulkheads, decks, and other structures?

___ Have you eliminated insulating foams, plastics, and other materials that produce toxic gases when burned?

Good fire-prevention practices will eliminate the need to test your fire-fighting skills.

Additional Information

Good information on fire extinguishers and fire fighting are available from:

National Fire Protection Association
Batterymarch Park
Quincy, MA 02269

Vessel Safety Manual
North Pacific Fishing Vessel Owner's Association
1800 W. Emerson, Suite 101
Fishermen's Terminal
Seattle, WA 98119

Navigation and Vessel Inspection Circular (NVIC) 5-86
U.S. Coast Guard
2100 Second St. SW
Washington, DC 20593-0001

Safe Seamanship

Safe seamanship is not automatically bestowed upon you just because you own a vessel. It comes instead with knowledge, practice, and an attitude that prudently evaluates risks. Lack of basic knowledge is the main reason for a great many "accidents," especially in the small-boat fleet.

This chapter introduces the fundamentals of rules of the road, stability, anchoring, actions to take if your vessel is holed or breaks down, emergency pumps, the role of drug use in vessel accidents, float plans, and emergency radio use.

Rules of the Road

All vessels in Alaska are governed by the International Rules of the Road. You can keep your vessel off a collision course by knowing these rules, and remembering that it may be necessary to deviate from the rules in order to avoid immediate danger.

Lookout

It is illegal for all hands to sleep while a vessel is underway or drifting. You must maintain a proper lookout at all times by sight and hearing, as well as by all available means appropriate (including radar). This constant watch allows you to fully appraise the situation and the risk of collision.

Safe Speed

You must proceed at a safe speed at all times so you can take action to avoid collisions, and can stop within a distance appropriate for the prevailing circumstances and conditions.

Avoid Collision

The Zones of Approach define your action and relationship to another vessel. The vessel that gives way should take early and substantial action to keep well clear. As the vessel with the right of way, it is your duty to see that a collision is avoided. As such, maintain course and speed, but be prepared to act if the give-way vessel does not take appropriate action. Risk of a collision exists if another vessel's compass bearing doesn't change or changes very little relative to you. When in doubt, assume the risk of collision exists.

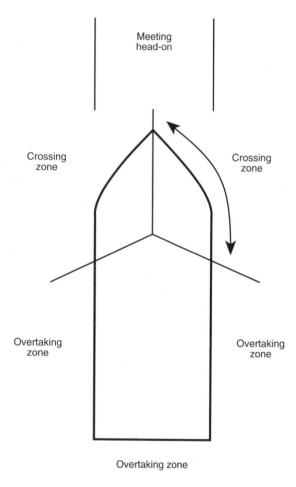

Zones of Approach.

Help avoid collisions by: not passing ahead of another vessel, taking early and positive action, making obvious course and speed changes, and slowing, stopping, or reversing if necessary.

In general, a vessel being **overtaken** has the right of way. When meeting **head-on**, vessels should pass port to port unless they have clearly communicated other intentions to one another over the radio.

In a **crossing** situation, if the other vessel is on your starboard side, **it** has the right of way. Keep out of the way and, as far as possible, avoid crossing ahead of the vessel.

When in doubt whether a situation is crossing or head-on, assume it's head-on and act accordingly.

Right of Way

The pecking order between vessels is:

1) Vessel not under command
2) Vessel restricted in her ability to maneuver
3) Vessel constrained by her draft
4) Vessel engaged in fishing
5) Sailing vessel
6) Overtaken vessel
7) Power-driven vessel
8) Seaplane

Give way to vessels above you on this list. In a narrow channel, vessels restricted to the channel have priority over fishing vessels, sailing vessels, and vessels under 20 meters.

Vessels fishing with trolling gear are **not** categorized as "vessels engaged in fishing" in the pecking order because their maneuverability is not considered restrictive.

Sound Signals

Horn blasts help you signal your intentions. One short blast means, "I am altering my course to starboard." Two short blasts mean, "I am altering my course to port." Three short blasts mean, "I am operating astern propulsion."

When there is danger or doubt as to who is doing what, the signal is five or more short blasts.

It is sometimes very difficult to hear another vessel's horn. In these cases, radio communication with the other vessel may be necessary to determine its intention.

Other sound signals for overtaking in a narrow channel, approaching an obscured bend, and making way in restricted visibility are detailed in the complete rules listing. The address for the International Rules of the Road publication is in the resources section.

Navigation Lights

Navigation lights can tell you a great deal about what nearby boats are doing. Learn to identify the lights you are most likely to encounter.

*This daymarker indicates channel location
in Kodiak, Alaska. (K. Byers photo)*

Buoys

All of North and South America use the same buoyage system to laterally mark areas safe for navigation. Buoys have standard shapes, colors, numbering, and position. Although buoys can be invaluable

Buoys

Returning from Sea[1]	Color	Number	Unlighted Buoy Shapes[2]	Light Color[3]	Daymark Shape[4]
Right side of channel	Red	Even	Nun	Red	Triangular
Left side of channel	Green	Odd	Can	Green	Square
Mid-channel	Red & white vertical stripes	May be lettered	Nun or Can	White	Octagonal

[1]Or heading in a northerly direction along the Pacific Coast, clockwise around a significant land mass, or in the general direction of flood tide.

[2]If the buoy is unlighted, it will have this shape. The small figure next to the buoy shows how that buoy would be marked on a chart.

[3]If the buoy has a light, instead of a specific shape, the light on the buoy will be this color.

[4]A daymark, a small marker held up by two poles and visible only during daylight hours, may also be present.

aids to navigation, their lights may extinguish, they can shift location or capsize, or they may be carried away by the tide, ice, or a vessel. If buoys don't appear to be where the chart says they should be, they may not be there. Beware!

The above table compares the different buoy characteristics.

Stability

Vessel instability leads to many accidents and deaths in the fishing industry. You can reduce the chance of capsizing by knowing the factors that affect stability and taking steps to ensure that your vessel remains stable enough for your operating conditions.

The term stability refers to the ability of a vessel to return to the upright position after being heeled by an external force. Do not assume that a comfortable ride indicates a stable ship. Vessels with a long period of roll from side to side may ride better than those that return upright quickly, but they can be unstable.

Stability can be viewed as the relationship between a vessel's buoyancy and its load. The load can be people, gear, supplies, and fish, or unplanned weight such as water on deck or spray ice. Factors that change a vessel's buoyancy or load, or cause the load to shift, can lead to a change in stability that can be life-threatening.

Stability problems often result from a combination of factors and frequently involve a change in weather or sea conditions. Some common vessel stability problems are listed below.

Conversions

Converting a vessel from one fishery to another can make a previously stable vessel very unstable. Many deaths could have been prevented if vessel owners had consulted with a naval architect about conversions.

The most dangerous type of conversion involves the addition of deck equipment, especially up high, and the addition or modification of deckhouses. As a general rule of thumb, you should have a stability analysis performed if the sum of the added weight and the removed weight total more than 1 percent of the original lightship weight (weight of the empty vessel).

Free Surface Effect

Shifting liquids or semi-liquids can cause a decrease in stability known as free surface effect. The shifting can be from any liquid or fish carried aboard, and can be in tanks, compartments, or on deck.

Free surface effect problems can be avoided by following your stability test recommendations and a few simple guidelines:

- Use checkers to keep loose fish from sliding around on deck, but make sure the on-deck water can drain. Keep the scuppers clear.

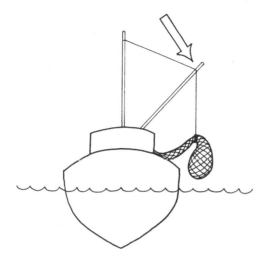

Lifting weight can make a vessel unstable. When an object is lifted, its weight is immediately transferred to the head of the lifting device.

- Keep crab or refrigerated seawater tanks either full or empty. Half-full or slack tanks are an invitation for a roll-over.

- Prevent the water level from dropping in saltwater tanks by installing a check valve in the discharge line, and install alarms that will tell you when tanks are not completely full or completely empty.

- In fish holds, use bin boards that run the length of the vessel. This can greatly reduce loss of stability due to free surface effect. Make sure the bin boards are secure and strong. The fish in the hold must **not** be allowed to shift.

Lifting Weight

Lifting a suspended weight can create problems, especially for marginally stable vessels. When an object is lifted, its weight is immediately transferred to the head of the lifting device. This can cause a vessel to heel past its range of stability.

Be aware of the dangers that lifting can cause. Ensure that your lifting devices are mounted no higher than necessary to be effective. Be careful if your gear is hung up and you are trying to pull it free. This can have the same effect as lifting a weight, and could cause your vessel to capsize.

Load Height

Generally speaking, a vessel is more stable when weight is stored as low as possible.

Stacking crab pots too high or carrying fish on deck can create a serious stability problem, and has caused scores of fishing deaths. Sadly, many of these capsized vessels had a stability report that specified the number of pots that could be carried safely, but the recommendations were ignored.

Environmental Factors

Certain sea conditions and other environmental factors can reduce your vessel's stability. Beam seas can be dangerous because they pour large amounts of water on deck, and can create free surface problems if the water isn't drained quickly.

Large following seas are especially problematic and should be avoided because of the resultant corkscrew-type rolling, the tendency to lose the ability to steer the vessel when the rudder and propeller are out of the water, and the propensity for the vessel to broach. Icing can also create serious stability problems. These are addressed in detail in the Weather chapter.

Watertight Integrity

Much of a vessel's stability rests on its ability to be buoyant, and buoyancy demands that dry spaces stay dry. Keep your vessel's flush hatches secure at all times. In addition, regularly clean the hatch's seating surface and check the gasket to make sure it will seat properly.

In rough weather, keep doors and hatches closed and secured to prevent water from entering the wrong places. A dutch door to the galley can permit good ventilation but keep water out when the weather is tough.

Install bilge alarms and check the bilge on a regular basis. Bilge alarms are cheap insurance.

Overloading and Freeboard

Overloaded vessels are often dangerously unstable and are compromised in several ways. Because the extra cargo or fish is usually carried on deck, the vessel's center of gravity is raised. This slows the roll, and generally decreases the tendency of a boat to return to its normal upright position.

Second, the deck load can begin to shift around, resulting in a free surface effect that will destabilize the vessel. Third, the extra weight sets the vessel deeper in the water, causing a decrease in the amount of freeboard. This is not a good situation.

Freeboard is critical because it represents the reserve buoyancy in a hull. With low or no freeboard, the ability of a ship to return to the upright position becomes severely compromised. Further, low freeboard means more waves coming on board. This means additional weight on deck, free surface effect problems, and even less freeboard than before.

Keep all of this in mind when tempted to carry those extra pounds of fish or cargo. Never exceed the loading conditions of your stability report. If your vessel does not have a stability booklet, be conservative and prudent in all your loading habits.

See the resources section in this book for references with further information about the more technical aspects of stability and the importance of stability testing.

Watches

Be sure your wheel watch understands what he's doing before he stands watch. As a minimum he should know basic rules of the road, whistle signals, radio distress signals, standard running lights, how to read a fathometer, and when to call the captain. Try to pair a greenhorn with an experienced hand on the same watch, and give orders to get someone to take over if they become sleepy.

Have a compass deviation table made up and posted, and be sure all hands know how to apply it.

Be sure you understand the effect of current on your boat. If you don't know what the currents are for your area, look them up in a Current Table. When the weather or current makes passage of a particular stretch of water so difficult you only **think** you can make it, remember that the good seaman admits his own limitations and those of his boat, and only does what he **knows** he can do safely.

Keep a dead-reckoning plot on a chart whenever underway in fog, even if you have every available piece of modern, red-hot electronic gear. Also, consider keeping a logbook with entries every 15–30 minutes showing time, location, speed, direction, etc. This will give you excellent backup in case of electronic failure.

Anchoring

If you want to sleep soundly and the holding ground is good, let out cable five to seven times the depth of water. If you expect some adverse weather, let out ten times the depth of water.

If you're anchoring in tough weather and want to improve your ride, lash three or four (or more) inflated buoys together and seize them to the anchor cable after you have the proper scope out. Then continue to release cable until the buoys are a couple of fathoms under water. **Dog** your winch, and you're guaranteed a better ride.

Set up a watch to check the anchor regularly for dragging, and set the alarm on your depth sounder.

Groundings

It is difficult for a fishing vessel to occupy a space in the water already filled by a rock. Although some rocks may not be charted properly, the vast majority of those that have been struck **are**. Most of the time a boat grounds because "somebody goofed."

Avoidance is the best defense against grounding, and for that you need good charts and full attention to navigation. Try out difficult passages for the first time in daylight and with a rising tide, go slowly, and post a lookout on the bow.

If you do end up aground, call the Coast Guard and do not attempt to refloat until you have inspected the damage as best you can. If you are taking on water, call the Coast Guard again. You may need to beach your vessel until either you can repair the damage or assistance arrives.

Holed Vessel

If your boat is holed with a large opening, call for assistance **immediately**, and head for the beach. It's usually a losing proposition to effectively plug large holes while afloat. Small holes can be plugged with almost any material on hand, including a wooden plug wrapped in cloth. Keep several aboard your vessel.

Another good method is to place a pillow over the hole, hold it tight, and brace it by any handy means. A mattress or canvas tarp secured at each outside corner can be dragged under the vessel to cover the hole, too. Although outside patches are often extremely ef-

fective, they are subject to chafing away and slippage. Get to a safe harbor or beach fast.

If your vessel is holed near the waterline and is set up so you can cause a deliberate list—perhaps by swinging out a heavy object on the end of a boom or trolling pole, or transfering fuel from one side to another—consider doing this on the side opposite the hole to reduce the pressure at the damaged spot. Do **not** do this if the list will make your vessel dangerously unstable.

Lost Propeller or Other Total Breakdown

Occasionally vessels lose their propeller or steering, or suffer engine breakdown. When this occurs, the vessel is suddenly plunged into a totally helpless situation and immediate assistance should be requested. The Coast Guard much prefers to be told "No further assistance required" when halfway there, than to lose the head start on a case.

If you have a total breakdown, anchor if you can. Often, however, the water will be too deep. If you are in open water far from land, and you don't have a commercially made sea anchor, make one. Anything that drags under water is a sea anchor—even a bucket or your regular anchor. When streamed over the bow on about 10 fathoms of line, a good sea anchor reduces the drift due to wind, and the boat will ride well, holding its head out of the trough.

Weather permitting, you may be able to tow the boat, or at least somewhat control the direction of drift, if you have a skiff.

Echoes

Echoes can be used during close inshore navigation to determine approximate distance to land. It takes about 10 seconds for sound to travel one mile and then return as an echo. The following table shows the relationship between the seconds to hear an echo and the distance away in nautical miles.

2 seconds - .2 miles
4 seconds - .4 miles
6 seconds - .5 miles
8 seconds - .7 miles
10 seconds - .9 miles
12 seconds - 1.1 miles
14 seconds - 1.3 miles
16 seconds - 1.4 miles
18 seconds - 1.6 miles
20 seconds - 1.8 miles

Operating Coast Guard Emergency Pumps

If you neglected to carry an emergency pump, or your pumps cannot keep up with incoming water, the Coast Guard can provide you with emergency, gasoline-powered pumps. The type you are most likely to get comes sealed in a rectangular orange plastic container about 3 feet high, and is capable of pumping about 120 gallons per minute.

USCG drop pump instructions.

1. Attach

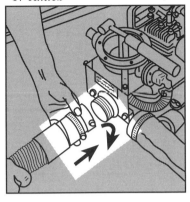

2. Immerse

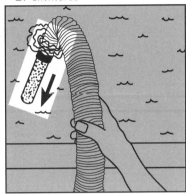

3. Overboard

4. Do not remove!

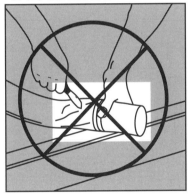

5. Attach

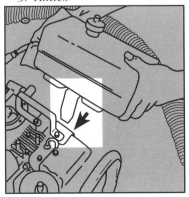

6. Open vent

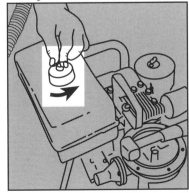

USCG drop pump instructions. (continued)

7. Attach
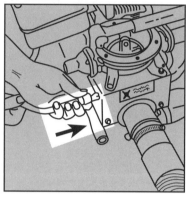

8. Prime
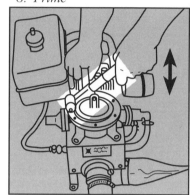

9. Choke
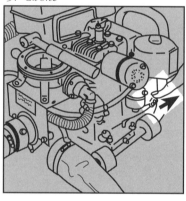

10. Start
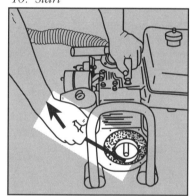

11. Run

12. Stop

One single lever on top opens it up, but don't smoke—there may be gasoline fumes inside the can. Make sure you keep the canister upright when removing the pump, or the oil may drain from the pump's crankcase, and the engine may seize on starting. There will be a waterproof flashlight and instructions in the canister.

Make sure you follow the instructions, or the pump will not work. Do **not** cut off the end of the discharge hose; this will prevent the pump from priming itself. Make sure all O-rings are in place.

Drugs and Fishing

Substance abuse is a problem within the fishing industry, just as it is in all walks of life. One of the dangerous things about alcohol and fishing is that alcohol is much more potent on board, drink for drink, than when it is consumed on land. This is due to the hypnotic effect of navigating on the water, the background of steady engine noise, and the long hours and limited sleep associated with fishing.

The effects of drinking alcohol have been clearly demonstrated by skippers who volunteered for an experiment that measured their ability to operate a boat under the influence. Despite the fact that all of the volunteers felt that their performance actually improved, both observers and objective measurements indicated the skippers' performance decreased. Not only did the skippers perform poorly, but their judgment of their own condition was faulty after they consumed alcohol.

Alcohol is not the only drug that causes problems in the fishing industry. Amphetamines, cocaine, and other drugs are all too common. Some can produce a rush that allows you to stay awake for hours on end, but they are highly addictive and can cause a decrease in judgment.

Who wants a crew member or skipper with impaired judgment and balance? Drugs and safe fishing just don't mix.

Float Plans

The Coast Guard frequently receives reports of overdue fishing vessels. Unfortunately, the party making the report often hasn't the faintest idea where the boat is. Alaska is a big place, and time and money are wasted when search areas are not clearly defined. If you were overdue, would anyone on shore or in another boat know where you are?

Filing a float plan can save time in an emergency—and may help save your life. You can file in person or by mail, but make sure the person you tell will miss you and initiate a search if you fail to reach your destination on time. Be sure to relay:

- Your vessel's name, number, and description.
- The names of people on board.
- Safety and survival equipment on board.
- Where you're headed.
- When and where you expect to return to port.
- Who to contact if you fail to return.

Radio Use in an Emergency

Having and using marine radios is an integral part of fishing and an invaluable aid in an emergency. It is also a privilege granted by the agency that issues the licenses—the Federal Communications Commission (FCC).

Emergency Frequencies

Emergency marine radio calls are made on VHF channel 16 (156.8 mHz) or SSB 2182 kHz. The Coast Guard and other vessels also monitor 4125 kHz.

Emergency Calls

There are three internationally recognized radio signals used for marine emergencies: MAYDAY, PAN-PAN, and SECURITY. All three have priority over other radio traffic.

Mayday calls also have priority over all other emergency signals. They are to be used **only** when a vessel or life is threatened by grave and imminent danger, and a request is made for immediate assistance. To transmit a Mayday:

Make sure your radio is on and you transmit on channel 16 VHF or 2182 kHz SSB or 4125 kHz SSB. Then state:

1) Mayday, Mayday, Mayday.
2) Your vessel name and call sign **three** times.
3) Position (latitude and longitude, and loran are preferred).
4) Nature of distress (fire, grounding, medical emergency, etc.).

5) Total number of people on board (P.O.B.).

6) Amount and type of survival gear on board (immersion suits, life rafts, EPIRB, flares, etc.).

7) Vessel description (length, color, type, etc.).

8) Listen for a response. If there is none, repeat the message until it is acknowledged or you are forced to abandon ship.

If time permits, provide the Coast Guard with any additional information they request. They are often unable to begin a search until they have specific details about the nature of the emergency.

If you hear a Mayday call and it is not answered, you must answer it and log the details of the call. When you can be reasonably sure you will not interfere with other distress-related communications, advise the vessel in distress what assistance you can offer.

MAYDAY RELAY: All vessels that are required to have radios (such as fishing vessels) are required to relay Maydays that are heard but go unanswered. To relay an unanswered Mayday:

Make sure your radio is on and you transmit on channel 16 VHF or 2182 kHz SSB or 4125 kHz SSB. Then state:

1) Mayday relay, Mayday relay, Mayday relay.

2) **Your** vessel's name and call sign.

3) Name and call sign of vessel in distress.

4) Location of vessel in distress.

5) Nature of problem with vessel in distress.

6) Degree of assistance needed.

7) Listen for acknowledgment.

8) Transmit additional requested information.

PAN-PAN (pronounced pahn-pahn) calls are for very urgent messages concerning the safety of a boat or persons. Examples include urgent storm warnings by an authorized station, and loss of steering or power in a shipping lane.

To transmit a PAN-PAN message:

Make sure your radio is on and you transmit on channel 16 VHF or 2182 kHz SSB or 4125 kHz SSB. Then state:

1) PAN-PAN, PAN-PAN, PAN-PAN all stations.

2) Your vessel name and call sign **three** times.

3) Nature of urgent message.

4) Position (latitude and longitude, and loran are preferred).

5)   Total number of people on board.

6)   Vessel description (length, color, type, etc.).

SECURITY (pronounced say-cure-i-tay) calls are the lowest priority emergency calls and are used to alert vessel operators to turn to another station to receive a safety message. SECURITY warns nearby vessels of a possible hazard.

Putting It All Together

There's a lot more to fishing than just knowing where the fish are and how to catch them. With lives, vessels, and livelihoods at stake, it pays to take the extra time to be prepared and to know what you're doing.

Orientation, Emergency Instructions, and Drills

It was around 6 a.m. on September 22, and the 70-foot wooden longliner F/V *Majestic* was heading for a halibut opening in the Bering Sea. Although the weather wasn't calm, the roll to starboard that didn't return was the crew's first indication of a problem. The vessel never did right itself.

In the seven frantic minutes between the time the *Majestic* heeled over and sank, the five men onboard fought 35-knot winds, rough seas and darkness as they tried to broadcast a Mayday, don immersion suits, get the EPIRB, and launch the life raft.

Things don't always go perfectly when you abandon ship, and they didn't for the crew of the *Majestic*. No one answered their Mayday, the raft caught in the rigging, one man was floating by himself with the EPIRB, and the other four were drifting in pairs, wondering where the EPIRB was.

All five were rescued after six hours in the water. In a subsequent interview, Skipper Tom Bedell said, "I'd like to stress for all fishermen: Have your drills. I'll bet you a lot of fishermen don't even know how to get into a survival suit and have never had drills.

"We had stations for everybody, and I explained it when we left port because there was a little bit of a crew change. Everybody had the full knowledge of what their station was, and that's exactly what everybody did. We've had drills before with [the former skipper]. He'd wake us up when we were just sitting around watching TV, or in the nighttime: 'The boat's going down!'

"Everybody'd run up to their stations and he'd clock us.

"I've been fishing for twenty years and I've been on quite a few boats. I've got to admit in a situation like that a lot of people would not know what to do. When the person has the full knowledge of what they're supposed to do before a crisis situation, you don't have to sit and communicate; everybody just does it. Had we not known our stations, we probably would have lost a few guys. It would have just been pandemonium." (*Alaska Fisherman's Journal*, November 1992.)

Many skippers **do** conduct onboard drills because they know that practice can make a difference. Others hold drills because they are required for their vessel. Mandated or not, the truth is: Drills help improve your chances of surviving an emergency at sea.

There are three components to a good drill:

1) Orienting all crew members to the vessel.

2) Giving crew members instructions that detail what they should do in the event of an emergency.

3) Conducting the actual drill.

What Is Required?

The Commercial Fishing Industry Vessel Safety Act (CFIVSA) of 1988 mandates a variety of equipment and training for certain fishermen. This chapter concentrates on orientation, emergency instructions, and drills, **which are required for documented commercial fishing vessels that operate beyond the boundary line with any size crew, and documented commercial fishing vessels with more than 16 individuals on board.**

These are the items that need to be practiced in drills at least once a month. All of these items are included in the four ready-made drills on pages 208-215.

- Abandon vessel.
- Fight a fire in different locations.
- Recover a person from the water.
- Minimize the effects of flooding.
- Launch and recover survival craft.
- Put on immersion suits and PFDs.
- Put on a fire fighter's outfit and self-contained breathing apparatus (SCBA) if so equipped.
- Make a radio distress call.
- Use visual distress signals.
- Activate the general alarm.
- Report inoperative alarm and fire detection systems.

After September 1, 1994, persons conducting the drills must be:

1) A USCG-accepted instructor OR

2) An individual licensed for operation of inspected vessels of 100 gross tons or more.

See the resources section of this book for a listing of training agencies. For a list of USCG-accepted instructors in your area, contact your local USCG unit and ask for the Officer in Charge of Marine Inspection (OCMI).

Some portions of the Act pertain to **all** commercial fishing vessels. For more specific details on the Act and what applies to you, contact the U.S. Coast Guard.

Orientation

Who's got time to orient a newcomer, especially one who has years of fishing experience? There's too much to do before a trip. And why bother telling your nephew Jay how to operate the radio? He'll never use it, he's only along for the ride.

Hold those thoughts. That new deckhand may not know how to turn on your vessel's EPIRB, even though he's no greenhorn. And your nephew might just need to use the radio if you get hurt.

Orienting people to your vessel has three steps:

- **Think** about what you want newcomers to know about your vessel's basic operation. Where are the safety hazards? Do they know how to operate the safety and survival gear? Would they know what to do in the event of an emergency? The list on page 188 will give you a good start on what to cover.

- **Show** them around the vessel. This will take time, but it will pay dividends in several ways: Crew members who are oriented to safety hazards are less likely to injure themselves, they will be able to help during an emergency, and you will greatly reduce your liability in the event of an emergency.

- Have them **sign the safety orientation log** (sample on page 218). Consider making a copy of this log and keeping it in a safe place at home in the event your vessel sinks. This record is very important for your legal protection.

Vessel Safety Orientation

Show Vessel Layout

- Engine: on/off, steering, gear selection, etc.
- Shut off and crossover valves.
- Alarms: what they are, what they mean, reporting inoperative alarms.
- Entrapment: exit routes.
- Hazards: hatches, winches, machinery, lines, slippery areas, stability concerns, etc.

Show Vessel Safety and Survival Equipment

- Immersion suit/PFD: need, stowage, fit, donning.
- Life raft/survival craft: need, location, function, deployment.
- EPIRB: need, location, function, deployment.
- Radio(s): need, location, function, use.
- Electronic position-fixing devices: function, how to find position.
- Flares: need, location, function, use.
- Fire extinguishers: location, function, use.
- Other survival equipment: line thrower, person overboard recovery gear, first aid kit, etc.

Show Vessel's Policies and Emergency Instructions

- Drug and alcohol policy.
- Placards: report all injuries, report all malfunctions, waste disposal.
- Emergency instruction: both posted and in book.

Emergency Assignments (Station Bill): Each Crew Member's Specific Duties In

- Abandoning the vessel.
- Fighting fires in different locations onboard the vessel.
- Recovering an individual from the water.
- Minimizing the effects of unintentional flooding.
- Launching survival craft and recovering lifeboats and rescue boats.

- Donning immersion suits and wearable PFDs.
- Donning fireman's outfit and SCBA (if so equipped).
- Making a voice radio distress call.
- Using visual distress signals.
- Activating the general alarm.
- Reporting inoperative alarm systems and fire detection systems.

The next several pages contain material that will help you conduct an effective orientation. This information also meets the requirements for emergency instructions that are mandated for certain vessels by the CFIVSA.

Show newcomers around the vessel so they will be familiar with its layout and can find safety and survival equipment if an emergency situation arises.

Emergency Instructions

According to the Act, the following vessels must have emergency instructions on board:

- Documented commercial fishing vessels that operate beyond the boundary line.
- Documented commercial fishing vessels with more than 16 people on board.

The emergency instructions must include:

- Emergency Equipment Location and Abandon Ship Stations (page 192).
- Emergency Assignments (Station Bill) and Signals (page 194).
- Distress Broadcast (page 196).
- Donning Immersion Suits (page 197).
- Anchoring Instructions (page 199).
- Person Overboard (page 199).
- Unintentional Flooding/Rough Weather at Sea/Crossing Hazardous Bars (page 200).
- Fire (page 201).
- Abandon Ship (page 202).

1) If you operate with **less** than 4 people on board, the instructions need to be readily available.

2) If you operate with **4 or more** people on board, some instructions must be posted and the others kept readily available.

Must Be Posted

- Emergency Equipment Location and Abandon Ship Stations (page 192).
- Emergency Assignments (Station Bill) and Signals (page 194).
- Distress Broadcast (page 196).
- Donning Immersion Suits (page 197).

Must Be Kept Readily Available

- Anchoring Instructions (page 199).
- Person Overboard (page 199).
- Unintentional Flooding/Rough Weather at Sea/Crossing Hazardous Bars (page 200).

- Fire (page 201).
- Abandon Ship (page 202).

If your vessel is not required to have these instructions onboard, that doesn't mean you don't need them. In fact, they could help save your life. Or your crew. Or your vessel. Don't assume that everyone knows what to do in an emergency. Think of these instructions as your opportunity to make sure everyone does what they are supposed to do.

Research on emergencies has shown that **the most important element in surviving is the person's initial reaction to the emergency and his training.** (Dr. Lars Weiseth, Norway)

General Instructions

F/V _____

Captain _____

1) **All** persons on board shall:
   a.   Report to the captain for an orientation briefing.
   b.   Have emergency duties.
   c.   Be responsible for knowing their emergency duties.
2) Emergency duties include knowing:
   a.   The location of lifesaving and emergency equipment.
   b.   How to operate assigned equipment.
   c.   How to make a distress call.
   d.   What to do in the event of a person overboard.
   e.   What to do in the event of a fire.
   f.   What to do in the event of an order to abandon ship.

Emergency Equipment and Abandon Ship Station

F/V _____

ABSTA	Abandon Ship Station	**O**	Life Rings
EPIRB	EPIRB	**PFD**	PFDs
FEX	Fire Extinguishers	**RADIO**	Radio
FL	Flares	**RAFT**	Raft
ISUIT	Immersion Suits		

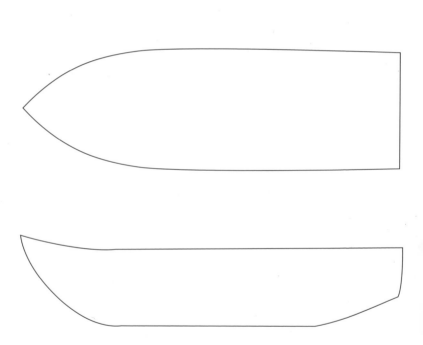

Emergency Equipment and Abandon Ship Station

F/V _____ *I'm A Wake* _____

ABSTA	Abandon Ship Station	**O**	Life Rings
EPIRB	EPIRB	**PFD**	PFDs
FEX	Fire Extinguishers	**RADIO**	Radio
FL	Flares	**RAFT**	Raft
ISUIT	Immersion Suits		

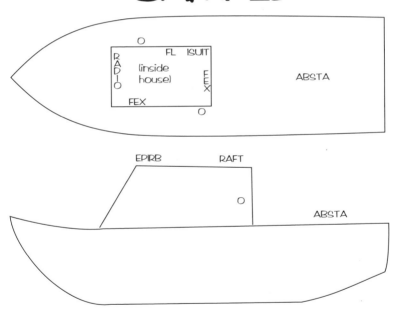

Emergency Assignments and Signals

Note: Alternate signals may be determined and practiced by vessel operator

Position	Person Overboard Signal: 3 blasts for the letter "O" (Morse code) repeated at least 4 times	Fire Signal: 1 long continuous blast not less than 10 seconds	Flooding Signal: 1 long continuous blast not less than 10 seconds	Abandon Ship Signal: At least 7 short blasts followed by 1 long blast not less than 10 seconds
	Station/Bring/Duty	Station/Bring/Duty	Station/Bring/Duty	Station/Bring/Duty
Captain				
All Others				

Emergency Assignments and Signals   SAMPLE

Note: Alternate signals may be determined and practiced by vessel operator	Person Overboard Signal: 3 blasts for the letter "O" (Morse code) repeated at least 4 times	Fire Signal: 1 long continuous blast not less than 10 seconds	Flooding Signal: 1 long continuous blast not less than 10 seconds	Abandon Ship Signal: At least 7 short blasts followed by 1 long blast not less than 10 seconds
Position	**Station/Bring/Duty**	**Station/Bring/Duty**	**Station/Bring/Duty**	**Station/Bring/Duty**
Captain	Wheelhouse, radio, maneuver vessel.	Wheelhouse, radio, maneuver vessel.	Wheelhouse, radio, maneuver vessel.	Wheelhouse, radio, maneuver vessel.
1st Deckhand	Throw flotation, lookout	Fight fire	Plug hole, pump	Immersion suits, life raft
2nd Deckhand	Don immersion suit/safety line	Assist in fire fighting	Assist in plugging hole & pumping	Immersion suits, life raft
Cook	Communicate, assist where needed	Communicate, boundary person, remove hazards, survival gear	Communicate, assist where needed, secure hatches	Communicate, EPIRB, count crew.
All Others				

Distress Broadcast

- Make sure your radio is on and you transmit on VHF channel 16 or 2182 kHz SSB or 4125 kHz SSB. Press microphone button. Speak slowly, clearly, and calmly. Say:

  MAYDAY, MAYDAY, MAYDAY.

  This is the F/V _____, F/V _____, F/V _____, call sign _____. Over.

- Release microphone button briefly and listen for acknowledgment. If no one answers, say:

  MAYDAY, MAYDAY, MAYDAY.

  This is the F/V _____, F/V _____, F/V _____, call sign _____.

  My position is _____. (Use latitude/longitude, nearby landmarks, distance from known points, LORAN readings, etc.) Repeat three times.

  I am _____. (sinking, on fire, listing, etc.)

  I estimate that I can stay afloat _____ hours/minutes.

  I have _____ persons onboard.

  My vessel is a _____ type of vessel, _____ feet long, has a _____ color hull with _____ color trim and _____ masts.

  I will be listening on channel _____.

  This is the F/V _____, call sign _____.

- If situation permits, stand by the radio to await further communications with the Coast Guard or another vessel.

- If no answer and situation permits, try another channel and repeat.

Donning Immersion Suits

Your life may depend on your ability to quickly don an immersion suit in an emergency, so it makes sense to have done it before. Monthly practice should reduce your donning time from minutes to seconds.

Vessel movement or list often prevents donning while standing, so practice donning the suit while sitting on the deck. Avoid diesel and other substances that might harm the suit.

Sit on the deck and work your legs into the suit, leaving boots or shoes on if possible. Placing plastic bags over your boots or shoes may make suit donning easier. Wear or bring extra warm clothing if possible.

Pull the hood over your head, then place one arm into each sleeve of the suit and reset the hood on your head. OR, place your weaker arm into the sleeve of the suit. Then reach up and pull the hood over your head with your free hand. Then place your strong arm into the suit.

Sit down and work your legs into the immersion suit.

Place your weaker arm into the sleeve of the suit first.

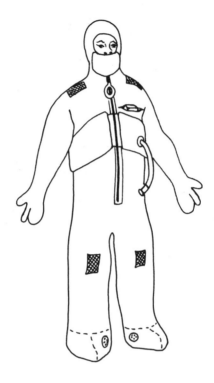

Hold the zipper below the slide with one hand, and fully close the zipper by pulling on the lanyard. Then secure the flap over your mouth.

Warnings

- Wearing an immersion suit inside a vessel may hamper your escape during a sudden capsizing.

- To prevent possible injury, do not inflate the air bladder until you are in the water.

- Ease or lower yourself into the water. Jump only if necessary.

- Beware of getting snagged on gear.

- **Keep the hood on**—it could save your life!

Anchoring Instructions

1) **Choose a location** with protection from the wind and seas, if possible, and with a suitable bottom.

2) Anchor in a **maximum water depth** of _____ feet or _____ fathoms. (Note: This vessel has _____ feet or _____ fathoms of line/chain.)

3) **Approach the anchorage location slowly**, and head the bow into the wind or current, whichever is stronger.

4) **When the vessel starts to back slowly, ease the anchor to the bottom.**

5) **Quickly pay out a scope** of 5 (in good weather) to 10 (bad weather) times the water depth in anchor line/chain. **Secure the anchor line/chain.** If drift is not rapid, back down with minimum power to set the anchor.

6) **Maintain an anchor watch** to feel the anchor drag and look out for any vessel drift. **Be prepared to get underway if vessel drags anchor.**

Person Overboard

Signal: 3 blasts for the letter "O" (Morse code) repeated at least 4 times

1) **Throw** a life ring buoy or flotation as close to the individual as possible.

2) Post a lookout to **keep the individual in the water in sight** and **communicate the distress and position to the pilothouse**.

3) Pilothouse watch to **sound alarm and maneuver as necessary**. Mark position.

4) Launch a **rescue boat or platform** to recover the individual **if appropriate**.

5) Have a **crew member put on an immersion suit**, attach a **safety line** to the crew member, and have crew member stand by to enter the water to **assist** in recovery **if appropriate**.

6) **If** individual overboard is **not immediately located, notify the Coast Guard and other vessels** in the vicinity and **continue searching** until released by the Coast Guard.

Unintentional Flooding, Rough Weather at Sea, Crossing Hazardous Bars

Signal: 1 long continuous blast not less than 10 seconds

1) **If unintentional flooding: Notify pilothouse immediately to sound alarm and call Mayday.**

   If rough weather at sea or crossing a hazardous bar is anticipated, notify the entire crew.

2) Close all watertight and weathertight doors, hatches, ports, and air vents to **prevent taking water aboard and further flooding** in vessel.

3) Keep bilges dry to **prevent loss of stability due to water** in bilges. Use power-driven bilge pumps, hand pumps, fire pumps, and buckets to dewater.

4) **Check all intake and discharge lines that penetrate the hull for leakage.** All crew members should know the location and operation of all through-hull fittings.

5) If on small vessel, crew should remain evenly distributed on the vessel.

6) Personnel should **don immersion suits/PFDs** if the going becomes very rough, the vessel is about to cross a hazardous bar, or **when** otherwise **instructed by the master or individual in charge of the vessel.**

Warning: Wearing an immersion suit or PFD inside a vessel may hamper your escape during a sudden capsizing.

Fire

Signal: 1 long continuous blast not less than 10 seconds

1) **Notify pilothouse immediately** to **sound alarm and call Mayday**.

2) **Shut off air supply to fire:** Close hatches, ports, doors, vents, etc.

3) **De-energize electrical systems supplying the affected space**, if possible.

4) **Assemble portable fire fighting equipment.**

5) **Account for personnel** and **fight fire**. Do not use water on electrical fires.

6) **If fire is in machinery space, shut off fuel supply** and use fixed extinguishing system if appropriate.

7) Maneuver vessel to **minimize effect of wind** on the fire.

8) **Move survival gear that could be damaged by fire.**

9) Check adjoining spaces to prevent spread of fire.

10) Once fire is extinguished, **begin dewatering to avoid stability problems.**

11) **If unable to control fire, notify Coast Guard and nearby vessels.** Prepare to abandon ship.

Abandon Ship

Signal: At least 7 short blasts followed by 1 long blast of 10 seconds or more

1) **Preparation** should include the following as time and circumstances permit:

   a. **General alarm and Mayday.**

   b. **All personnel don immersion suits/PFDs** and warm clothing.

   c. **Prepare to launch life raft.** Attach sea painter to vessel above weak link.

   d. **Get abandon ship kit,** including **signals** (**EPIRB**, flares, smokes, flashlights, handheld radios, etc.), **first aid kit, water,** and **food**.

   e. Gather other useful items.

2) **Meet** at abandon ship station.

3) **When sinking is imminent** or remaining on board is inappropriate, close watertight openings and:

   a. **Launch and board life raft.**

   b. **Keep raft's sea painter attached to the vessel but be prepared to cut sea painter** immediately if there is risk of damage to raft or vessel sinks.

   c. **Activate EPIRB and begin 7 Steps to Survival.**

Drills

Once all newcomers have been oriented to your vessel and know their emergency assignments, you need to do monthly drills whenever the vessel is in service. Sure, the CFIVSA drill requirement only pertains to documented commercial fishing vessels operating beyond the boundary line or documented commercial fishing vessels with more than 16 people on board, but drills save lives.

The crew onboard the 56-foot *Angela Marie* was anxiously waiting for the water to clear from the back deck. When it didn't, the crew knew it was time to don immersion suits. That's not easy in 60- to 80-knot winds when your vessel is lying on its side and its surface is icy.

There were more problems. The Mayday had to be rebroadcast as the vessel sank because people hadn't clearly understood their position, one man went in the water without his suit, and the skipper struggled to untangle himself from lines in the water.

Then good things began to happen: The life raft surfaced next to the crewman with no suit and he was able to climb in. The Coast Guard arrived on scene and hoisted all five men to safety.

What did the skipper have to say afterward? The crew "all did what they were supposed to do. Thank God for their experience and their levelheadedness. With any panic at all, it could have been a different situation. The best advice I can give anybody is to do your drills." (*Alaska Fisherman's Journal*, April 1994)

How to Conduct Effective Drills

Drills don't have to be boring, and they shouldn't be exactly the same each month because emergencies vary. Here are some clues to make your vessel's drills more effective:

Be Realistic

- Realism will help make the drills more interesting. Instead of saying, "Okay, let's have a drill on abandoning ship," think of a place, time of day, weather and sea conditions that you might find yourself in when you have to abandon ship. Then describe that to the crew and position them where they would likely be. Or, ask the crew to describe a circumstance they think they might encounter. Make the scenarios as realistic as possible.

- Some actions, such as inflating the life raft during the abandon ship drill, should be simulated. Drills will be more effective if crew members say what they **would** do if it were a real emergency. It will be hard for some crew to pretend, but a realistic scenario will help.

- Crews should feel some pressure; pressure is normal. But all crew members must understand that **their personal responsibility is to be safe**. It is part of the job.

Be Spontaneous

- Don't always announce drills ahead of time, but always announce a drill as a drill.

- Conduct drills at different times and places. Try them at the dock, underway, at night, in the rain. Be creative!

Make drill scenarios as realistic as possible.

Do Hands-on Drills

- Retention for hands-on learning is 90 percent, much higher than just talking about or watching a videotape of a drill.

- Be sure the crew is familiar with the vessel's emergency instructions, signals, and assignments (station bill) as well as the location and use of survival equipment.

- Have crew members touch the equipment as much as is practical. Have them put on their immersion suits, actually back each other up in a fire scenario, say a Mayday (without the radio on), etc.

- **Disconnect the radio's microphone** or make other provisions to prevent a false Mayday.

- Stress familiarity with equipment during both day and night.

- Avoid damaging actual emergency equipment.

Make Drills Progressive

- Start with simple walk-throughs and build skill and speed, but **never include running**.

- Progress to more complicated scenarios.

- Throw in "curves" to make scenarios more interesting. For example, you might want to have one crew member simulate being nearly debilitated by seasickness or injury to determine whether people will help him, and to ensure that others assume his duties.

Build Teamwork

- Teamwork increases efficiency and saves lives. Forget the "every man for himself" philosophy. It has sabotaged the saving of lives and ships at sea.

- Build on the team that you already have for fishing.

- Cross-train crew members to cover each others' responsibilities in the event of an injury or other circumstance. If the skipper is injured, is the crew prepared?

- Be sure **all** hands participate in the drills.

Each crew member should talk about what was learned and how the drill could be done better.

Be Positive

- Drills can be fun—a chance to feel good about those you must count on in an emergency.
- Drills should **not** be used to punish, harass, intimidate, or frustrate your crew.

Debrief All Drills

- **A drill is not over until it is debriefed.**
- Each crew member should talk about what was learned and how the drill could be done better. Everyone should feel that it is okay to make constructive comments on the vessel's and crew's performance. Drills should be a positive experience for everyone.
- Copies of the vessel's Emergency Assignments and Signals (page 195) and Emergency Instructions (page 190) will help keep the discussion on track, as will a list of each drill's critical points (see pages 208–215 for four suggested drills).

- Consider changes to Emergency Assignments and vessel's emergency equipment. Remember, the largest room in the world is the room for improvement.

- Inspect and return all gear to its proper location after each drill so it is ready and available for a real emergency.

Log the Drill

- Complete written log entry describing date and location of the drill, and evaluation of crew and individual responses. There's a log sheet for drills on page 216 and one for testing your EPIRB on page 217.

Ready-made Drills

The following four drills, which contain all the items required by the CFIVSA (page 186), have been devised to help fishermen and other mariners conduct their own emergency drills. Running time for each is about 20 minutes. Both novice and experienced drill leaders may discover ideas and critical points to look for while conducting and evaluating drills. Conduct these drills once a month and you will be in compliance with the federal onboard drill regulations.

Skippers will find it convenient to use these prepared drills. If you choose to develop your own drills be sure they cover all 11 drill items on page 186.

Person Overboard

Scenario: While hauling gear during sloppy weather, a deck hand is washed overboard by a large wave or falls overboard while dipping a 5-gallon bucket over the side. The crew member is wearing a flotation suit equipped with a light and whistle. Other boats are visible in the area.

Before drill: Be sure the crew is familiar with the vessel's person overboard recovery plan, including:

___ How the skipper plans to pull a person back onboard.

___ What equipment is required.

___ Skipper's requirements for wearing flotation while on deck.

___ Rules for being on deck in rough weather or at night.

___ Setting up the drill

This drill is best run while underway with no gear in the water, **and with the person overboard represented by an inflated buoy** with a personal marker light attached.

Initiating the drill: The drill leader chooses a "victim" and informs that crew member about the overboard incident. The drill leader then **throws a buoy overboard** and advises another crew member of the person overboard. The "victim" does not participate in the drill directly, but observes the crew's reactions to the scenario and helps keep track of the following critical points.

Critical points to look for during drill:

Alarms/Communication

___ Does person discovering the emergency initiate the alarm?

___ Does person discovering the emergency tell the wheelhouse which side of the vessel the victim fell off?

___ Does person on watch alert all crew members? How?

___ Are Coast Guard and other vessels made aware of the problem?

___ Does entire crew recognize the Man Overboard signal?

___ How soon is entire crew aware of the emergency?

___ Are any crew members unaware of the emergency due to an inoperative signal or lack of communication?

___ Is communication to the wheelhouse sufficient to bring the vessel to the victim?

___ Does crew communicate with each other?

___ Is simulated distress signal called off after the victim is rescued?

Response

___ Do crew members react in accordance with their Emergency Assignments?

___ Do crew members readily do unassigned but needed jobs (cross-trained)?

___ Does crew work together as a team?

___ Do crew members anticipate or react to events?

___ Does person discovering the emergency throw a marker?

___ Does person discovering the emergency continually keep the victim in sight and point?

___ Does person on watch use electronic position fixing devices to mark the position of the person overboard?

___ Does person on watch initiate a proper maneuver?

___ How long does it take to rig the recovery device?

___ Is crew in place, including a rescue swimmer in an immersion suit with a safety line, by the time the vessel is back alongside the victim?

___ Is recovery device and vessel's hauling equipment used effectively?

___ Do any crew members endanger themselves by leaning perilously over the side to recover the victim?

___ Does crew recognize hypothermia and know appropriate treatment for the victim?

___ Is medical help sought for treatment of hypothermia, if needed?

___ Is crew aware of considerations in recovering survival craft?

Fire On Board

Scenario: Fire is caused by a faulty diesel stove, clothing placed too close to an electrical heater, frayed insulation on electrical wiring against a bulkhead, an oversized light bulb in a bunk light, or other appropriate cause.

Setting up the drill: This drill is best run while underway, at the beginning of a trip, and with no gear in the water. Fire can be simulated by strobe lights or a red rag. Tape can be used to block off passages due to "smoke." This drill can easily evolve into an abandon ship drill.

Initiating the drill: Tell a crew member that there is smoke and/or flames coming out of the stove/state room/bulkhead. When the crew member is clear on how to correctly report the fire, the drill begins. Keep the drill moving by telling the crew how effectively they are controlling the fire as the drill proceeds. The fire can spread or be extinguished, depending on their efforts.

Critical points to look for during drill:

Alarms/Communication

___ Does person discovering the fire immediately sound the alarm?

___ Does person on watch alert all crew members? How?

___ Are Coast Guard and other vessels made aware of the problem?

___ Does entire crew recognize the Fire signal?

___ How soon is entire crew aware of the emergency?

___ Are any crew members unaware of the emergency due to an inoperative signal or lack of communication?

___ Does crew report information such as source and size of fire, and number of persons involved?

___ Is communication to the wheelhouse sufficient to allow operator to maneuver vessel to minimize the effect of wind on the fire?

___ Does crew communicate with each other?

___ Do crew members account for others?

___ Is simulated distress signal called off once fire is under control?

Response

___ Do crew members react in accordance with their Emergency Assignments?

___ Do crew members readily do unassigned but needed jobs (cross-trained)?

___ Does crew work together as a team?

___ Do crew members anticipate or react to events?

___ Does operator maneuver the vessel to minimize effect of wind on the fire?

___ Is operator safely able to leave the wheel, if necessary, to inspect the affected area?

___ If help is not available, does operator close doors and seal openings to isolate the fire?

___ Are areas near the fire that are vented or have operating machinery or fans closed or secured?

___ Are electricity and fuel sources to the affected space secured?

___ Do crew members go around, rather than pass through, smoke-filled spaces when evacuating the affected area?

___ When evacuating affected areas, do crew members remove portable extinguishers, immersion suits and other survival equipment, and hazardous items?

___ If an installed fire suppression system is used, is it only activated on word from the skipper, and only after vents, doors, and hatches are secured and all persons evacuated?

___ Do firefighters don Self-Contained Breathing Apparatus (if equipped) or fight fire by staying low?

___ Are firefighters always backed up?

___ Is an appropriate extinguishing agent used?

___ Do crew members act as if they are familiar with extinguisher advantages, disadvantages, and range?

___ Is fire or deck hose brought to the scene and pumps placed in line?

___ Do crew members act as if they are aware of the hazards of toxic smoke and gases?

___ Are fire boundaries checked periodically to prevent the fire from spreading?

___ If the fire is not controlled, are initial preparations made to abandon ship?

___ If water is used to control the fire, are provisions made to dewater the vessel?

___ How is it determined that the fire has been extinguished?

___ Once the fire is out, is a reflash watch set and the affected areas overhauled?

Flooding

Scenario: The vessel is running from the fishing grounds with a deck load of fish and gear. Wind and seas are rising and are off your quarter.

Before drill: Make sure the crew is familiar with the vessel's plumbing system, through-hull fittings, pumps, and equipment available for damage control.

Setting up the drill: This drill can be run any time and can evolve into an abandon ship drill. The drill leader will inform the crew of the location of the "flooding" and the level of the water.

Initiating the drill: The drill leader tells crew members that the vessel seems to be getting sluggish, and asks them to check lazarettes, holds, and the engine room. The drill leader then informs the crew of the location and extent of the problem. Keep the drill moving by telling the crew the level of flooding. Let them know how effectively they are controlling the problem as the drill proceeds.

Critical points to look for during drill:

Alarms/Communication

___ Does the person discovering the emergency initiate the alarm?

___ Does crew report information such as location, extent, and cause of flooding?

___ Does person on watch alert all crew members? How?

___ Are Coast Guard and other vessels made aware of the problem?

___ Does entire crew recognize the General and High Water alarms?

___ How soon is the entire crew aware of the emergency?

___ Are any crew members unaware of the emergency due to an inoperative signal or lack of communication?

___ Is communication to the wheelhouse sufficient to maneuver the vessel to lessen risk of capsizing?

___ Does crew communicate with each other?

___ Do crew members account for others?

___ Is simulated distress signal called off once the flooding is under control?

Response

___ Do crew members react in accordance with their Emergency Assignments?

___ Do crew members readily do unassigned but needed jobs (cross-trained)?

___ Does crew work together as a team?

___ Do crew members anticipate or react to events?

___ Does person on watch initiate appropriate maneuvers to lessen risk of capsizing: Reduce speed? Head into seas? Minimize roll?

___ Is person on watch safely able to leave the wheel, if necessary, to inspect the flooded area?

___ What actions are taken to improve stability?

   ___ Fish/gear tossed?

   ___ Freeing ports cleared?

   ___ Free surface effect minimized?

   ___ Blocks lowered?

   ___ Cross-flooding minimized?

   ___ Stability plan used?

___ Is watertight integrity maintained by closing all watertight doors, hatches, etc.?

___ Are through-hull fittings, shaft housings, and other penetrations checked for leakage?

___ Is everyone familiar with operation of the vessel's pumps?

___ Are tarps, plugs, blankets, etc. used to slow leaks?

___ Are extra pumps (hand and power) and buckets used to dewater?

___ If gas pumps are used below decks, are CO/CO_2 problems considered?

___ Are there problems with the vessel's pumps?

___ Do crew members prepare survival equipment (life rafts, immersion suits, EPIRBs, extra clothing, water, food, flares, log, first aid kit, etc.) in case of sudden loss?

Abandon Ship

Scenario: Despite the crew's best efforts to control the fire or the flooding, the situation gets out of control and the drill leader gives the order to abandon ship.

Setting up the drill: This drill can be added to the end of a fire or flooding drill to save time and make the drills more challenging. To prevent the fire or flooding drill from being cut short, the drill leader should tell the crew not to abandon ship until the order is given. The crew will only **simulate** launching life rafts, activating EPIRBs, and abandoning the vessel. However, immersion suits should be donned and appropriate survival equipment brought to the abandon ship station.

Initiating the drill: When the fire or flooding drill has been concluded, the abandon ship signal will be sounded over the ship's alarm system.

Critical points to look for during drill:

Alarms/Communication

___ Does person on watch alert all crew members? How?

___ Are Coast Guard and other vessels made aware of the problem?

___ Does entire crew recognize the Abandon Ship signal?

___ How soon is entire crew aware of the emergency?

___ Are any crew members unaware of the emergency due to an inoperative signal or lack of communication?

___ Does the crew communicate with each other?

___ Are all crew members accounted for?

___ Are signals used or simulated before abandoning ship to attract nearby assistance?

___ Are all crew members able to make an adequate Mayday call and find the vessel's position?

Response

___ Do crew members react in accordance with their Emergency Assignments?

___ Do crew members readily do unassigned but needed jobs (cross-trained)?

___ Does crew work together as a team?

___ Do crew members anticipate or react to events?

___ Do crew members know their abandon ship station?

___ Do obstructions block escape routes or access to survival equipment?

___ Is life raft painter **always** secured (simulated) once life raft is released?

___ Do all hands have an immersion suit of a size that fits appropriately even with deck clothing on?

___ Do all crew members completely don their immersion suits in 60 seconds?

___ Does crew use a buddy system in donning suits and launching rafts?

___ Does crew simulate tossing throwable flotation (buoys, etc.) overboard?

___ Does crew gather an EPIRB, extra clothing, water, food, flares, log, and any other survival equipment, and are these items protected from washing overboard?

___ Are watertight doors and hatches closed, if there is time, before abandoning vessel?

___ Do crew members simulate entering the water properly wearing immersion suits?

___ Can all crew members describe how and when to launch a life raft and entry procedures?

___ Is crew aware of procedures for recovering a life raft?

___ Are all crew members aware of immersion suit features, proper care and stowage?

___ Can all crew members describe how to operate and test EPIRBs?

___ Is the EPIRB tested and logged at the end of the drill?

___ If flares are lit, is a Security given on channel 16?

Monthly Drills Log

Drill Performed	Date	Date	Date	Date	Date	Date	Date	Date	Date	Date	Date	Date
Abandoning vessel												
Fighting a fire in different locations												
Person overboard												
Minimizing the effects of flooding												
Launching survival craft & recovering life boats & rescue boats												
Donning immersion suits/PFDs												
Donning fire fighting outfit & SCBA (if so equipped)												
Mayday and using visual distress signals												
Activating the general alarm												
Reporting inoperative alarms and fires												

Monthly EPIRB Test Log

F/V _____

Battery Expiration Date _____

_____

_____

Hydrostatic Release Expiration Date _____

_____

_____

Date	Time	Comments	Date	Time	Comments

Note: EPIRBs should be tested monthly. The 406 mHz EPIRBs can be tested any time. All other EPIRBs should be tested for only 1-2 seconds during the first 5 minutes of any hour. Batteries and hydrostatic releases are dated and should be changed as indicated by manufacturer.

Vessel Safety Orientation Log

F/V _____

   This certifies that I have read and received a safety orientation briefing on this vessel, including the instructions, emergency assignments, and diagrams contained herein, and that I understand the above.

Date	Name (print)	Signature

Resources

Accidents

Accident Reporting Requirements

1) Commercial fishermen are required by law to file Coast Guard form 2692, Report of Marine Accident, Injury, or Death, when the following occur:

   a. All accidental groundings and any intentional groundings that also meet any of the other reporting criteria or create a hazard to navigation, the environment, or the safety of the vessel.

   b. Loss of main propulsion or primary steering, or an associated component or control system, the loss of which causes a reduction of the maneuvering capabilities of the vessel. Loss means that systems, component parts, subsystems, or control systems do not perform the specified or required function.

   c. An occurrence materially and adversely affecting the vessel's seaworthiness or fitness for service or route, including but not limited to fire, flooding, failure of or damage to fixed fire extinguishing systems, lifesaving equipment, or bilge pumping system.

   d. Loss of life.

   e. Injury that requires professional medical treatment beyond first aid, and, in the case of a person engaged or employed on board a vessel in commercial service, that renders the individual unfit to perform routine vessel duties.

   f. An occurrence not meeting any of the above criteria but resulting in damage to property in excess of $25,000. Damage cost includes the cost of labor and material to restore the property to the condition that existed prior to the casualty, but it does not include the cost of salvage, cleaning, gas freeing, drydocking, or demurrage.

A $1,000 civil penalty may be assessed if form 2692 is not filed within 5 days of the incident.

2) Commercial fishermen who are involved in a serious marine incident are required by law to be tested for both drug and alcohol use. The testing must take place at the earliest convenient time that does not hinder vessel safety. A serious marine incident includes the following events involving a vessel in commercial service:

   a. Any marine casualty or accident that results in any of the following:

   - One or more deaths.

   - An injury to the crew member, passenger, or other person that requires professional medical treatment beyond first aid, and, in the case of a person employed on board a vessel in commercial service, that renders the individual unfit to perform routine vessel duties.

   - Damage to property in excess of $100,000.

   - Actual or constructive total loss of any vessel subject to inspection under 46 U.S.C. 3301.

   - Actual or constructive total loss of any self-propelled vessel, not subject to inspection under 46 U.S.C. 3301, of 100 gross tons or more.

   b. A discharge of oil of 10,000 gallons or more into the navigable waters of the United States, as defined in 33 U.S.C. 1321, whether or not resulting from a marine casualty.

   c. A discharge of a reportable quantity of a hazardous substance into the navigable waters of the United States, or a release of a reportable quantity of a hazardous substance into the environment of the United States, whether or not resulting from a marine casualty.

National Transportation Safety Board (NTSB) Reports

The NTSB has the authority to investigate aircraft, train, bus, and fishing vessel accidents. It usually limits its marine investigations to accidents involving ships of at least 100 gross tons valued at more than $500,000, or when six or more lives are lost. It also holds hearings when accidents of a potentially recurring nature occur, or if there is widespread public interest. NTSB has investigated several fishing vessel accidents in the last decade. Copies of their reports, which also contain a section on recommendations that might help prevent similar accidents, may be ordered from:

National Technical Information Service
5285 Port Royal Road
Springfield, VA 22161
(703) 487-4630

Books

Interesting Reading about Sea Survival

Desperate Journeys, Abandoned Souls, by Edward Leslie. This is an excellent collection of stories about survivors from historical times to the present, and it provides many insights into the psychological aspects of survival. Engrossing reading.

Endurance: Shackleton's Incredible Voyage, by Alfred Lansing. This story of Ernest Shackleton's ill-fated expedition to reach the South Pole in 1914 is particularly insightful in terms of the vital role a leader can play in a survival situation. One of the all-time great survival stories.

Shackleton's Boat Journey, by F.A. Worsley. Although this book describes the same Shackleton expedition as *Endurance*, it does so from the captain's perspective.

Survivor, by Michael Greenwald, edited by Steve Callahan and Dougal Robertson. This is probably one of the more comprehensive books ever done on sea survival and is full of true stories. The 25 pages of sea bird identification may be a bit overdone, but no topic relating to sea survival has been left out.

Wild, Edible Foods References

Wild, Edible and Poisonous Plants of Alaska. Available from the Cooperative Extension Service, University of Alaska.

Surviving on the Foods and Water from Alaska's Southern Shores, Marine Advisory Bulletin No. 38. Available from the Alaska Sea Grant College Program, University of Alaska Fairbanks, address on title page of this book.

Recommended References to Have on Board

Make sure all references are up-to-date. Many of the following references are required onboard certain vessels by the Commercial Fishing Industry Vessel Safety Act of 1988.

- **Charts**
- **Coast Pilot, Volume 8 and 9**
- **Light List** (Coast Guard Publication COMDTPUB #P16502.6)
  Superintendent of Documents
  U.S. Government Printing Office
  Washington, DC 20402
  (888) 293-6498   gpoaccess@gpo.gov
- **Icing Report** (NOAA Data Report PMEL-14)
  NOAA/PMEL
  7600 Sand Point Way NE
  Seattle, WA 98115-0070
  (206) 526-6235
- **Local Notice to Mariners**
  Commander
  Aids to Navigation Branch
  17th Coast Guard District
  P.O. Box 25517
  Juneau, AK 99802-5517
  (907) 463-2133
- **Navigation Rules International-Inland**
  Superintendent of Documents
  U.S. Government Printing Office
  Washington, DC 20402
  (888) 293-6498   gpoaccess@gpo.gov
- **Beating the Odds on the North Pacific: A Guide to Fishing Safety (this book)**
- **Tidal Current Tables**
- **Tide Tables**
- **Vessel Safety Manual**
  North Pacific Fishing Vessel Owners Association (NPFVOA)
  1900 West Emerson, Suite 101
  Seattle, WA 98119
  (206) 285-3383

The NOAA Icing Report is a recommended reference to have on board your fishing vessel to help predict vessel icing rates. (USCG photo)

Boundary Lines and High Seas

The terms "high seas" and "boundary line" are both referenced in the Code of Federal Regulations (CFR) that affects commercial marine operation. Safety requirements for fishing vessels are determined by the waters they operate in as defined by these terms.

High seas are those waters **beyond** the territorial seas. (Territorial seas extend to three miles offshore.) The dividing line between territorial and high seas appears as a wavy line on standard NOAA navigation charts.

The **boundary lines** are defined in 46 CFR, Part 7. "Except as otherwise described in this part, Boundary Lines are lines drawn following the general trend of the seaward, highwater shorelines and lines continuing the general trend of the seaward, highwater shorelines across entrances to small bays, inlets and rivers." This general rule applies in Alaska for all areas except where the following lines define the boundary lines.

Canadian (BC) and United States (AK) Borders to Cape Spencer, AK

a. A line drawn from the northeasternmost extremity of Point Mansfield, Sitklan Island 40° true to the mainland.

b. A line drawn from the southeasternmost extremity of Island Point, Sitklan Island to the southernmost extremity of Garnet Point, Kanagunut Island; thence to Lord Rock Light; thence to Barren Island Light; thence to Cape Chacon Light; thence to Cape Muzon Light.

c. A line drawn from Point Cornwallis Light to Cape Bartolome Light; thence to Cape Edgecumbe Light; thence to the westernmost extremity of Cape Cross.

d. A line drawn from Surge Bay Entrance Light to Cape Spencer Light.

Cape Spencer, AK, to Cape St. Elias, AK

a. A line drawn from the westernmost extremity of Harbor Point to the southernmost extremity of LaChaussee Spit at Lituya Bay.

b. A line drawn from Ocean Cape Light to latitude 59°31.9′N longitude 139°57.1′W (Yakutat Bay Entrance Lighted Whistle Buoy "2"); thence to the southeasternmost extremity of Point Manby.

c. A line drawn from the northernmost extremity of Point Riou to the easternmost extremity of Icy Cape.

Point Whitshed, AK, to Aialik Cape, AK

a. A line drawn from the southernmost extremity of Point Whitshed to the easternmost extremity of Hinchinbrook Island.

b. A line drawn from Cape Hinchinbrook Light to Schooner Rock Light "1".

c. A line drawn from the southwesternmost extremity of Montague Island to Point Elrington Light; thence to the southernmost extremity of Cape Puget.

d. A line drawn from the southernmost extremity of Cape Resurrection to the Aialik Cape.

Kenai Peninsula, AK, to Kodiak Island, AK

a. A line drawn from the southernmost extremity of Kenai Penin-
    sula at longitude 151°44.0′W to East Amatuli Island Light; thence
    to the northwesternmost extremity of Shuyak Island at Party
    Cape; thence to the easternmost extremity of Cape Douglas.

b. A line drawn from the southernmost extremity of Pillar Cape on
    Afognak Island to Spruce Cape Light; thence to the easternmost
    extremity of Long Island; thence to the northeasternmost extrem-
    ity of Cape Chiniak.

c. A line drawn from Cape Nunilak at latitude 58°09.7′N to the
    northernmost extremity of Raspberry Island. A line drawn from
    the westernmost extremity of Raspberry Cape to the northern-
    most extremity of Miners Point.

Alaska Peninsula, AK, to Aleutian Islands, AK

a. A line drawn from the southernmost extremity of Cape Kumlium
    to the westernmost extremity of Nakchamik Island; thence to the
    easternmost extremity of Castle Cape at Chignik Bay.

b. A line drawn from Second Priest Rock to Ulakta Head Light at
    Iliuliuk Bay entrance.

c. A line drawn from Arch Rock to the northernmost extremity of
    Devilfish Point at Captains Bay.

d. A line drawn from the easternmost extremity of Lagoon Point to
    the northwesternmost extremity of Cape Kutuzof at Port Moller.

Alaska Peninsula, AK, to Nunivak, AK

a. A line drawn from the northernmost extremity of Goose Point at
    Egegik Bay to Protection Point.

b. A line drawn from the westernmost extremity of Kulukak Point
    to the northernmost extremity of Round Island; thence to the
    southernmost extremity of Hagemeister Island; thence to the
    southernmost extremity of Cape Peirce; thence to the southern-
    most extremity of Cape Newenham.

c. A line drawn from the church spire located in approximate posi-
    tion latitude 59°45′N longitude 161°55′W at the mouth of the
    Kanektok River to the southernmost extremity of Cape Avinof.

Kotzebue Sound, AK

A line drawn from Cape Espenberg Light to latitude 66°52'N longitude 163°28'W; and thence to Cape Krusenstern Light.

Communications

The Federal Communications Commission (FCC) and the Coast Guard on documented vessels requires you to have a ship's radio station and radio operator's license if you operate a ship's radio. License applications are available from many marine electronic dealers and Coast Guard offices, or from:

Federal Communications Commission
P.O. Box 1040
Gettysburg, PA 17325
(717) 337-1212

Fishing Permits

Alaska Department of Fish and Game

In addition to having permit applications, most local Fish and Game offices have copies of the regulations for all fisheries.

If you are commercially harvesting miscellaneous shellfish such as sea cucumbers or sea urchins, you need to contact the local or regional Alaska Department of Fish and Game biologist or the Division of Commercial Fisheries office in Juneau.

If you are interested in selling or processing fish or shellfish, contact the Division of Commercial Fisheries in Juneau. They can provide you with an "Intent to Operate" form and will direct you to the other state agencies involved in this permitting process. You may be required to post a bond with the Alaska Department of Labor, get a business license from the Department of Revenue, and be inspected by the Department of Environmental Conservation.

Division of Commercial Fisheries
Alaska Department of Fish and Game
P.O. Box 3-2000
Juneau, Alaska 99802-2000
(907) 465-4210 or (907) 465-4150

Commercial Fisheries Entry Commission

Permit cards are required for every commercial fishery in Alaska and are issued by the Commercial Fisheries Entry Commission offices in Juneau and Kodiak. Contact them or your local Alaska Department of Fish and Game office for an application form. Be sure to acquire other necessary permits or licenses required for your fishery, and to allow adequate time for your applications to be processed.

Commercial Fisheries Entry Commission
8800-109 Glacier Hwy.
Juneau, AK 99801
(907) 789-6150

Commercial Fisheries Entry Commission
c/o Alaska Department of Fish and Game
211 Mission Road
Kodiak, AK 99615
(907) 486-4791

*Fishermen loading halibut into a net bag at a Cordova,
Alaska, processing plant. Fishermen need two permits if
they are landing halibut in Alaska waters. (K. Byers photo)*

International Pacific Halibut Commission

If you are commercially fishing for or landing halibut in Alaska state waters, you need a permit card from the Commercial Fisheries Entry Commission, and a permit from the International Pacific Halibut Commission.

International Pacific Halibut Commission
P.O. Box 95009
Seattle, WA 98145-2009
(206) 634-1838

National Marine Fisheries Service

Anyone who commercially fishes for groundfish (except halibut) beyond the three mile limit must have a permit card from the Commercial Fisheries Entry Commission, and a federal groundfish permit from the National Marine Fisheries Service (NMFS).

Information on the Marine Mammal Protection Act, the Endangered Species Act, and the Lacey Act (which governs the marking of containers or packages that contain fish or game transported across state lines) is also available from NMFS.

National Marine Fisheries Service
P.O. Box 21668
Juneau, Alaska 99802-1668
(907) 586-7225

National Marine Fisheries Service
1211 Gibson Cove
Kodiak, Alaska 99615
(907) 486-3298

U.S. Department of Commerce
NOAA Office of Law Enforcement
329 Harbor Drive, Suite 210
Sitka, Alaska 99835
(907) 747-6940

Immersion Suit Repair

If your immersion suit needs repair, be sure to take or send it to an authorized repair shop. The suit's manufacturer should be able to recommend an approved facility. One such place is:

King Neptune
810 NW 45th St.
Seattle, WA 98107
(206) 783-5512
(800) 592-6255

Regulations

Contact one of the following U.S. Coast Guard offices for information on commercial fishing vessel regulations that address such things as vessel inspection, manning, required equipment, operating a motor vessel while intoxicated, sexual abuse reporting, marine pollution, zero tolerance, marine sanitation devices, drug testing, personnel licensing and misconduct, reporting accidents and deaths, and written agreements with crew members that detail the crew share and deducted expenses.

Commanding Officer
Marine Safety Office
2760 Sherwood Lane, Suite 2A
Juneau, AK 99801-8545
(907) 463-2450

Supervisor
Marine Safety Detachment
2030 Sea Level Drive #203
Ketchikan, AK 99901
(907) 225-4496

Supervisor
Marine Safety Detachment
329 Harbor Drive, Room 202
Sitka, AK 99835-7554
(907) 966-5454

Commanding Officer
Marine Safety Office
510 L Street, Suite 100
Anchorage, AK 99501-1946
(907) 271-6700

Supervisor
Marine Safety Detachment
c/o CG Support Center
P.O. Box 5A
Kodiak, AK 99619-0055
(907) 487-5750

Supervisor
Marine Safety Detachment
150 Trading Bay Rd., Suite #3
Kenai, AK 99611-7716
(907) 283-3292

Commanding Officer
USCG Marine Safety Office
P.O. Box 486
Valdez, AK 99686-0486
(907) 835-7210

Commander (mfvs)
Seventeenth Coast Guard District
P.O. Box 25517
Juneau, AK 99802-5517
(907) 463-2025

Training and Videos

First aid and CPR Training

For information on training in your area inquire at your local fire department or ambulance service, or contact:

Emergency Medical Services Section
Box 110616
Juneau, AK 99811-0616
(907) 465-3027

Marine Safety and Survival Training

Alaska Marine Safety Education Association (AMSEA)
Box 2592
Sitka, AK 99835
(907) 747-3287

Alaska Vocational Technical Center (AVTEC)
Box 889
Seward, AK 99664
(907) 224-3322

Ocean Safety Services
Box 2314
Homer, AK 99603
(907) 235-7908

Clatsop Community College
Marine Sciences Division
1653 Jerome Avenue
Astoria, OR 97103
(503) 325-0910 ext. 2380

Coast-Wise Marine Safety Training
Bev Noll
1385 Pebble Beach Drive
Crescent City, CA 95531
(707) 465-4400

Educational Training Company
Box 464
Sitka, AK 99835
(907) 747-5454

Marine safety and survival training, including immersion suits, is important for fishermen. Contact the agencies listed here for training schedule. (K. Byers photo)

Fremont Maritime Services
501 N. 36th Street, #217
Seattle, WA 98103
(206) 522-5377

Kodiak Community College/Kodiak Fishermen's Wives
117 Benny Benson Drive
Kodiak, AK 99615
(907) 486-4161

Mid-Atlantic Safety and Survival
P.O. Box 8453
Norfolk, VA 23503
(804) 661-3845

"Stability for Fishermen" Correspondence Course
National Cargo Bureau, Inc.
30 Vesey Street - Sixth Floor
New York, NY 10007

North Pacific Fishing Vessel Owners Association (NPFVOA)
1900 West Emerson, Suite 101
Seattle, WA 98119
(206) 285-3383

Marine and Industrial Safety Association
P.O. Box 1978
Port Isabel, TX 78578
(210) 943-7935

Vessel Safety Corporation
Paul Helland
P.O. Box 2075
Kingston, RI 02881
(800) 856-2394

Washington Sea Grant
    Ed Melvin
    1801 Roeder Avenue, #128
    Bellingham, WA 98225
    (206) 650-1527

    Sara Fisken
    1735 W. Thurman Street
    West Wall Building, #124
    Seattle, WA 98199
    (206) 543-1225

    Steve Harbell
    P.O. Box 88
    South Bend, WA 98586
    (206) 875-9331

Videos

The following videos are available for rent or purchase from most of the training agencies listed above. If they are not, contact the agency that produced the tape (listed in parentheses).

Cold Water Near Drowning, MAPV-2—Background and treatment of cold water near drowning. (MAP)

*Conducting Onboard Drills.**

*Emergency Radio Procedures.**

EPIRBs—How EPIRBs work, and are maintained and tested.*

Fire Prevention and Control—Preventing and fighting fires at sea. (NPFVOA)

Fishing Vessel Stability—General points about fishing vessel stability and stability testing. (NPFVOA)

Frostbite and Other Cold Injuries, MAPV-17—Recognition and treatment of frostbite and other cold injuries, but not hypothermia. (MAP)

Hypothermia, MAPV-1—Recognition and treatment of hypothermia. (MAP)

Immersion Suits—Donning, testing, inspecting, and stowing of immersion suits.*

Inflatable Life Rafts—Installing, launching, and boarding inflatable life rafts.*

It Could Have Been Prevented, MAPV-18—Boating safety for small boats with emphasis on the dangers of alcohol and boating. (AMSEA, MAP)

*Marine Fire First Response.**

Marine Survival Equipment and Maintenance, MAPV-11—Use and maintenance of marine safety equipment. (MAP)

Medical Emergencies at Sea—Treating medical emergencies at sea. (NPFVOA)

Personal Flotation Devices—Types of PFDs and how they work.*

Safety Equipment and Survival Procedures—Maydays, life raft launching, and survival kits. (NPFVOA)

Sea Survival, MAPV-3—Handling a sea survival emergency in a life raft with emphasis on the Seven Steps to Survival. (MAP)

Shore Survival, MAPV-4—Survival on shore with emphasis on the Seven Steps to Survival. (MAP)

Visual Distress Signals—The use of pyrotechnics, radios, and lights as emergency signals.*

*If these eight videos are not available from your nearest training agency, they may be purchased from:

    John Sabella & Associates
    805 W. Emerson St.
    Seattle, WA 98119
    (888) 719-4099

MAP videos are available from:

    Marine Advisory Program
    University of Alaska Fairbanks
    2221 E. Northern Lights Blvd., Suite 110
    Anchorage, AK 99508-4140
    (907) 274-9691

Useful Facts and Figures

1 fathom = 6 feet
120 fathoms = 1 cable
7½ cables = 1 statute mile
8.44 cables = 1 nautical mile
6,076 feet (about 2,000 yards) = 1 nautical mile = 1 minute of
    latitude

32 points of the compass = 360 degrees
1 point = 11.25 degrees

1 gallon of fresh water = 8.355 pounds
1 cubic foot = 7.481 gallons of water
1 cubic foot of fresh water = 62.5 pounds
1 cubic foot of salt water = 64 pounds
1 ton of fresh water = 35.84 cubic feet
1 ton of salt water = 35 cubic feet
261.8 gallons of sea water = 1 ton

1 cubic foot of normal storage ice = 62 pounds

1 barrel (42 gallons) of diesel = 315 pounds

1 gallon of engine motor oil = about 7.3 pounds

Doubling the diameter of a pipe increases the carrying capacity of
    the pipe four times.

Vessel Inspections

Courtesy Motor Boat Examinations/Dockside
Examinations

Courtesy Motor Boat Examinations are for recreational vessels
and are conducted by specially trained and qualified members of the
Coast Guard Auxiliary. Dockside Examinations are for commercial
fishing vessels and are conducted by Coast Guard personnel (military and civilian) and, in some cases, Auxiliary members. Examiners,
although generally not commercial fishermen, are qualified to inspect safety items common to all craft. If your boat passes the examination, you will be given a decal for your boat to indicate it has
passed. If you do not pass the examination, **no** report is made to
any law enforcement official.

Vessel Inspection and Maintenance References

Navigation and Inspection Circular No. 5-86
Commanding Officer
Marine Safety Center
U.S. Coast Guard
2100 Second St. SW
Washington, DC 20593-0001
attn: NVICs

Guide for Building and Classing Vessels
American Bureau of Shipping
45 Eisenhower Drive
P.O. Box 910
Paramus, NJ 07653-0910
(201) 368-9100

Index